Lecture Notes in Chemistry

Edited by G. Berthier, M. J. S. Dewar, H. Fischer,
K. Fukui, H. Hartmann, H. H. Jaffé, J. Jortner,
W. Kutzelnigg, K. Ruedenberg, E. Scrocco, W. Zeil

2

Enrico Clementi

Determination of
Liquid Water Structure
Coordination Numbers for Ions and
Solvation for Biological Molecules

Springer-Verlag
Berlin · Heidelberg · New York 1976

Author
Prof. Dr. Enrico Clementi
Istituto Guido Donegani
Società Montedison
Via del Lavoro, 4
I-28100 Novara

Library of Congress Cataloging in Publication Data

Clementi, Enrico.
 Determination of liquid water structure,
coordination numbers for ions, and solvation for
biological molecules.

 (Lecture notes in chemistry ; 2)
 Includes bibliographical references.
 1. Solvation. 2. Molecular association.
3. Water. I. Title.
QD543.C72   541'.342   76-41666

ISBN-13: 978-3-540-07870-8   e-ISBN-13: 978-3-642-93052-2
DOI: 10.1007/978-3-642-93052-2

DETERMINATION OF LIQUID WATER STRUCTURE, COORDINATION NUMBERS FOR IONS AND SOLVATION FOR BIOLOGICAL MOLECULES.

Enrico Clementi

Istituto Guido Donegani — Società Montedison — Via del Lavoro, 4 — Novara (Italy)

Foreword

The structure of liquid water and the solvation of ions and molecules represent an active field of past and current research; recently two reviews on "Structure of Liquids" have been presented by P. Schuster, W. Jakubetz and W. Marius and by S.A. Rice in Topics in Current Chemistry (Springer-Verlag, Berlin, Heidelberg, New York, 1970, volume 60) covering most of the recent progress up to about 1974. Both reviews have stressed that new important applications due to numerical simulations are in the making. In a way, this work represents the continuation of the previous two reviews; we have stressed in particular the new quantum mechanical developments that constitute the conceptual base for the recent advancements in this field. For this reason in the first part (concerning quantum mechanical techniques) we have pointed out, with a variety of examples, the large amount of information embodied in quantum mechanics that is often ignored either because not too important in discussing small molecular systems or because relatively too new.

To us the study of solvation represents the ideal field to pass from small chemical systems to large ones (and without symmetry), to pass from quantum mechanics (geometry, density and energy) to statistical mechanics (geometry, density, energy and temperature) with a first step towards thermodynamics (from quantized systems to continuous systems). Particular attention is given to present an unified picture that coherently retains technique and assumption in passing from one type of the description of matter to another.

Whereas in Part 1, we present the quantum mechanical bases and several novel techniques intended to be used for large systems, in Part 2 we consider in detail the problem of the structure of liquid water, a test case to see how reliable is the tool developed in Part 1. In

Part 3, we apply the proposed and tested tool to ion pairs, single ions, solvation around amino acids and other chemical systems of biological importance. Much of Part 3 represents very recent work (even the study of the structure of liquid water, our test case, is a very recent achievement). Therefore, we consider Part 3 simply as a small window on a vast horizon. This review is partly based on a volume by E. Clementi "V Coloquio International de Quimicas Teoricas de Expresion Latina; Revision de la Computacion Mecanico-Cuantica de Atomos y Moleculas" edited by the IBM Research Laboratory, San Jose, California, USA (July, 1974).

This work is dedicated to Prof. Per-Olov Löwdin, on the occasion of his 60th birthday. With this, we wish to stress that the quantum chemistry of large systems needs at its side the molecular physics of the small system, where new algebras and new models are proposed and tested and thus constitute the starting point for those extensions and adaptations necessary to deal with the large chemical systems (or more simply, with everyday chemistry).

Contents

Part 1   Description of a Chemical System as a Set of m Fixed Nuclei and n Electrons

1.1) <u>Introduction</u> - In order to be able to describe a molecule M surrounded by many molecules of water, one must be able to describe at
least one molecule of water interacting with M and one molecule of
water interacting with a second molecule of water (notice that the
simpler case is the one where M itself is a molecule of water). Implicit in the above statement is the assumption that the pairwise additivity approximation holds to a first approximation; implicit in
the title of the chapter is that we operate within the Born-Oppenhai-
mer approximation.
In this chapter we shall describe a theoretical framework where we
shall not have to be concerned with defining and differentiating between systems composed either of one or more molecules, or of molecules and ions, or of molecular complexes. All we shall have to state
are the nuclear positions - considered as fixed in space - the corresponding charges, and the number of electrons constituting the chemical system in consideration. The theoretical framework must provide
us with appropriate quantitative techniques that will univocally state
whether our system is formed by one or more molecules, or by atoms
or ions and molecules, or by molecular complexes, and must provide us
even with the best chemical structural formulas, so as to be in position to have a dialogue with traditional chemistry.
This task is accomplished by recurring to the LCAO-MO representation
(and its improved forms like SCF-LCAO-MO, C.I., MC-SCF, later described)
and to the analyses of the wavefunction (electron population analysis) and of the total energy (bond energy analysis). It is noted
that for a long time - till the middle 60 - molecular complexes have
been analyzed in the Valence Bond approximation[1], not in the LCAO-MO
approximation; then after a detailed study of the $NH_3+HCl \rightarrow NH_4Cl$
reaction, the situation changed[2]. Even today, often one finds in
literature the term "supermolecule approach" for those studies where
the SCF-LCAO-MO approximation is applied to two or more interacting
molecules. Likely, the main reason for these "esitations" must be
found in the conceptual difficulty to accept the equivalence of the
descriptions of a molecule in terms of atoms or in terms of nuclei
and electrons. In this sense the distinction between "basis set" - as
an analytical means to span the Hartree-Fock space - and "atomic orbitals" - as a physical entity - represents the conceptual division between the old quantum chemistry (up to 1960-1965) and the new quantum

chemistry (after 1960-1965).

In this work we shall not make use of semi-empirical techniques, but we shall deal only with ab-initio methods, because these: i) correspond to a more rigorous quantum-mechanical description of nature, ii) allow for a more rational and quantitative analysis of the errors due to the adopted approximations and iii) seem to represent the main stream for all past advances in quantum chemistry.

1.2) The SCF-LCAO-MO Approximation - A pleasant characteristic of the molecular orbital theory is that each progressive improvement, or step, has a natural and simple physical explanation.
Let us consider the LCAO-MO approximation[3]. There are actually two approximations implied in the above name: the first is the MO approximation, the second is the LCAO approximation to an MO. As known, the short notation LCAO-MO stands for "linear combination of atomic orbitals - molecular orbitals".
The MO is a one-electron function which is factored into a spatial component and a spin component. The expression "one-electron function" means that only the coordinates of one electron are explicitly used in a given MO[4]. This factorization into spatial and spin components is permissible, since generally one uses a hamiltonian which does not explicitly contain spin dependent terms. The MO's are the exact analog of the atomic orbitals, which describe the electrons in an atom to a first approximation. Indeed, one can read several chapters of the classical works of Condon and Shortly[5], replace the word "AO" with the word "MO" and one will read a book on molecular physics instead of atomic physics.
This situation has one important consequence: namely, a large amount of testing and development for molecular wavefunction techniques can be done with atoms.
If the molecule contains 2n electrons (let us consider a closed shell case for simplicity), the MO approximation will distribute the electrons in n molecular orbitals φ_1, φ_2,... φ_n. Since there are two possible spin orientations (α and β) a space distribution function has either spin α or β and, therefore, the 2n electron system is described by n spacefunctions and 2n spin-orbitals. Thus φ_1 and $\overline{\varphi}_1$ will have the same space distribution (will depend on the coordinates of one electron alone), but, in accordance with the Pauli exclusion principle, will have different spin functions. It is stressed that the one-electron model is acceptable because it simplifies the quantum mechanical treatment of 2n electrons. Indeed, since the very beginning of quantum

theory, Hylleraas introduced a wavefunction for the He atom in which one orbital is described in terms of the coordinates of both electrons. The total wavefunction Ψ_0 of the 2n electron system is then

$$
\Psi_0 \; = \; \frac{1}{\sqrt{(2n)!}}
\begin{vmatrix}
\varphi_1(1) & \cdots & \varphi_1(2n) \\
\overline{\varphi}_1(1) & \cdots & \overline{\varphi}_1(2n) \\
\cdot & & \cdot \\
\cdot & & \cdot \\
\cdot & & \cdot \\
\varphi_n(1) & \cdots & \varphi_n(2n)
\end{vmatrix}
\tag{1}
$$

where the number between parentheses indicates a given electron. This determinant wavefunction guarantees that any interchange of two electrons (i and j) brings about a sign change in the wavefunction. This is the Pauli constraint for fermions. The energy for such a system is given by the relation

$$
E_0 \; = \; \langle \Psi_0 | \, H \, | \Psi_0 \rangle
\tag{2}
$$

where the hamiltonian H is

$$
H \; = \; - \sum_i \tfrac{1}{2} \nabla_i^2 \; - \sum_{ia} \frac{Z_a}{r_i} \; + \sum_{ij} \frac{1}{r_{ij}} \; - \sum_{ab} \frac{Z_a Z_b}{R_{ab}}
\tag{3}
$$

The first term is the kinetic operator for the i-th electron, the second term is the potential operator between the i-th electron and the a-th nucleus (with charge Z_a), the third term is the electron-electron potential between the i-th and the j-th electrons, and finally the last term is the nucleus-nucleus potential with R_{ab} the distance between the a-th and b-th nucleus of respective charges Z_a and Z_b.
The first and second terms are subsequently referred to as the one-electron hamiltonian and will be indicated as h_0. The total energy for such a determinant was given by J.C. Slater, and it is

$$
E_0 \; = \; 2 \sum_i hi \; + \sum_{ij} (2 Jij - Kij) \; + \; ENN
\tag{4}
$$

where

$$
h_i \; = \; \langle \varphi_i | \, h_0 \, | \varphi_i \rangle
\tag{5}
$$

$$J_{ij} = \langle \varphi_i(1) \, \varphi_j(2) | r_{12}^{-1} | \varphi_i(1) \, \varphi_j(2) \rangle \tag{6}$$

$$K_{ij} = \langle \varphi_i(1) \, \varphi_j(2) | r_{12}^{-1} | \varphi_i(2) \, \varphi_j(1) \rangle \tag{7}$$

$$E_{NN} = \sum_{ab} (Z_a Z_b / R_{ab})$$

As known, the quantities J and K are usually referred to as coulomb and exchange terms, respectively.

What form should the MO have? Clearly, the molecular orbitals are subjected to symmetry constraints (as in the case of atomic orbitals) and any molecular orbital will transform as an irreducible representation of the molecular symmetry group. This statement, however, is not a sufficient one; indeed it tells us mainly how the molecular orbital should not be. In principle we could insist on the analogy between atomic one-electron functions and molecular one-electron functions and "tabulate" the MO in a way analogous to the method of Hartree and Fock in the 1930's. This would ensure that we have the best possible molecular orbitals. It is noted that numerical Hartree-Fock functions for diatomic molecules are a somewhat tempting possibility; this, however, has not seriously been explored at the present.

Nevertheless, chemistry is concerned with more than only diatomic molecules. An answer is provided by the LCAO approximation, in which the MO's are built up as linear combinations of atomic functions, namely

$$\varphi_i = \sum_j c_{ij} \chi_j \tag{8}$$

where χ_j in the old point of view are the atomic orbitals, and nowadays are simply one of the terms of a basis set.

We refer to R. S. Mulliken's classical series of papers for the early development and application of the LCAO MO approximation[3].

The next step in the evolution of quantum theory is the introduction of self-consistency. Again, the physical model is provided by atomic physics, namely by the Hartree-Fock model. The LCAO approximation to the MO requires the best possible linear combination: this is what one intends for self-consistency. A good review paper on this subject is the one by C.C.J. Roothaan[6]. There the self-consistent field technique in the LCAO MO approximation (SCF LCAO MO) is systematically exposed for the closed shell case.

Up to now we are strictly in the "one-electron approximation". The electrons interact among themselves only via the average field and the MO contains no explicit electron-electron parameter. Fortunately, the

Pauli principle keeps electrons with parallel spin (in different MO's) away from each other, but it has nothing to offer to electrons with antiparallel spin in the same MO. The full catastrophe might be appreciated by recalling that in the SCF LCAO MO approximation, two fluorine atoms are incapable of giving molecular bonding when brought together; i.e., the SCF LCAO MO does not recognize the existence of the F_2 molecule[7]. Of course, it does not require a computation of F_2 to realize this point. For example, when Roothaan's work appeared (1950), another less familiar paper was written by Fock[8] indicating how the problem could be solved and introducing the concept of two-electron molecular functions or "geminals". At the same time Lennard-Jones and collaborators[9] put forward a classical series of papers in which part of the correlation problem was tentatively solved, but at the expense of drastic orthogonality restrictions. For a variety of reasons, neither of the two avenues was numerically explored and in the meantime other techniques slowly emerged.

1.3) The Correlation Energy Correction - What is wrong with the wavefunction Ψ_o as given in Eq. (1)? The answer is available for example in the work by E. Wigner in 1934[10]. If we ask the question: what is the statistical relation between the position of two electrons, say electron 1 and 2, from the wavefunction (1), we have to solve for the integral

$$\int \Psi_o \, \delta(x-x_1) \, \delta(x-x_2) \, \Psi_o \, \prod_i (1,2) \, d\tau_i \tag{9}$$

where x stands for the cartesian and spin coordinates, and the integral is carried over all electrons coordinates except for those of electron 1 and 2 in consideration. The above integrals yield (by indicating with σ the spin variables and neglecting normalization)

$$\sum_{i=1}^{2n} \sum_{j=1}^{2n} \left[1 - \varphi_i(1)\varphi_i(2)\varphi_j(1)\varphi_j(2) \, d\sigma_i(1) \, d\sigma_i(2) \, d\sigma_j(1) \, d\sigma_j(2)\right] \tag{10}$$

For the case of two electrons with both parallel spin, the second term above is, in general, different from zero. For the case where the two electrons have opposite spin the second term of the equation above is zero. This means that the wave function (1) allows a pair of electrons with parallel spin to "feel" each other's relative position, to be correlated, whereas for the case of electrons with antiparallel spin the wavefunction allows any position with equal probability. Thus, the Hartree-Fock determinant introduces correlation in pairs of electrons with

parallel spin, but does not correlate those of antiparallel spin. Since
the second term of eq. (10) comes about because of the antisymmetry
of the wave functions, required for fermions, the second term's effect
is referred to as "Fermi hole". Thus, in the Hartree-Fock wavefunction
we have Fermi potential holes for the system of parallel spin, but no
hole between two electrons with antiparallel spin. Since clearly two
electrons with opposite spin should never occupy the same position simultaneously, the Hartree-Fock function lacks a mechanism to provide
for a potential hole experienced by an electron with spin α, when in
the neighborhood of an electron with spin β. The hole not present in
the wavefunction (1), the Hartree-Fock function, is referred to as
"Coulomb hole".

The energy gained by the system in introducing the Coulomb hole in the
wavefunction which represents such a system is the <u>correlation energy</u>.
This name was introduced by Wigner in the work previously quoted[10].
This energy gain, can be re-defined therefore as the energy difference
between the exact nonrelativistic energy and the Hartree-Fock energy[11].

The energy gained by the system in introducing the "Fermi hole" in the
wavefunction which represents such a system, is called <u>pre-correlation</u>
<u>energy</u>. This energy can be re-defined, therefore, as the energy differ-
ence between the Hartree and the Hartree-Fock energy[12].

To introduce the Fermi hole, the form of the wavefunction was changed
from a simple product (Hartree) to a determinant (Slater). To introduce
the Coulomb hole, we again have to change the form of the wavefunction. A large number of papers are devoted to this subject since the
late 1920's and we have no intention to review the extremely abundant
literature. We note that one possibility was put forward as early as
1930, namely to add corrections to the Hartree-Fock energy using wavefunctions of the form[10]

$$\Psi = \Psi_o + \lambda \phi \qquad (11)$$

where Ψ_o gives the Hartree-Fock energy and $\lambda \phi$, the correlation correction.

It is also noted that the total density distribution given by Hartree's
functions is not far different from the density distribution given by
Hartree-Fock functions. It is, therefore, reasonable to assume that the
density distribution variation needed for the introduction of a Coulomb
hole (relative to the Hartree-Fock density) is also small. Therefore,
we can use the Hartree-Fock density as a good approximation to the cor-

rect density and attempt to extract relations which should give the correlation energy from knowledge of the Hartree-Fock density. This task was solved by Wigner[10] who did propose a relation between density and correlation energy. We shall refer to this work as "statistical estimate of the correlation energy".

1.4) <u>Statistical Estimate of the Correlation Energy</u> - We shall now consider in detail the quantitative aspect of the correlation energy to atoms and then we shall discuss molecules.

The statistical model of Wigner has been revised a number of times in attempts to extend its validity from the cases where the density is high to cases of smaller density. Gombas[13] attempted to give an additional expression which covers both the high and low density region. His expression for the correlation energy, E_c, is as follows:

$$E_c = \int \rho^2 \, \varepsilon_c \, (\rho) \, d\rho =$$

$$= \int \rho^2 \, a_1 \rho^{1/3} \, (a_2 + \rho^{1/3})^{-1} d\rho \; + \; \int \rho^2 b_1 \, \ell n (1 + b_2 \rho^{1/3}) d\rho \tag{13}$$

where $a_1 = 0.0357$, $a_2 = 0.0562$, $b_1 = -0.311$, $b_2 = 2.39$ are Gombas' constants and ρ is the total density of the system.

It should be noted that if in the first term we put $a_2 = 0$, then clearly it contributes to the correlation energy E_c by an amount $a_2 N$ where N is the number of electrons. In Fig. 1 the value of the correlation correction using Hartree-Fock functions is given for the first and second row atoms.

If we consider the statistical model as a useful fitting formula then we can improve the situation.

Clementi and Salez[14] have used the following modified relation

$$E_c = a_1 \int \rho^{4/3} \, (0.0562 + \rho^{1/3})^{-1} \, d\rho \tag{14}$$

where a_1 is a numerical constant obtained from the relation (for an n electron system):

$$a_1 = 0.0237 + (0.0279n - 0.08176) \, (n + 3.45)^{-1} \tag{15}$$

In Fig. 1 the experimental correlation energies[15], and those obtained with the use of eqs. (13) and (14) are compared; one can conclude that these equations provide a reliable estimate of the correlation energy.

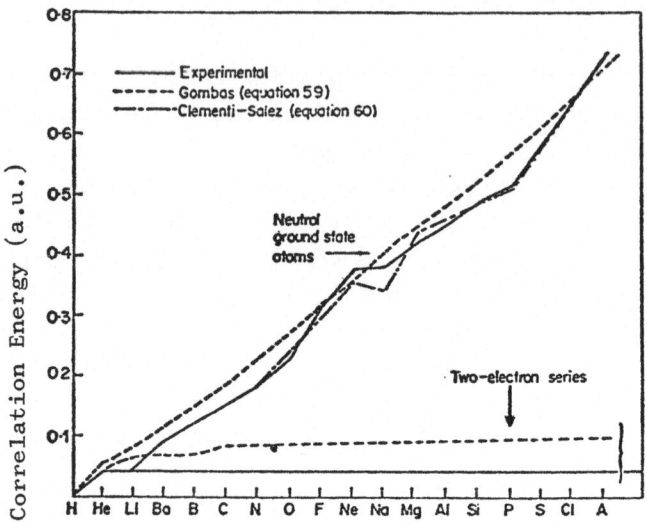

Figure 1. Correlation energy from statistical models. The solid line gives the experimental correlation energy for neutral atoms from helium to argon and for the isoelectronic series He, Li^+, Be^{2+},...A^{16+}. The dash-dash and the dash-line report the correlation energy using Gombas or Clementi and Salez relations. For the two electron isoelectronic series, the Clementi-Salez data are in exact agreement with the experimental one.

1.5) Configuration Interaction - Hylleraas[16] proposed the possibility of using not only one determinant, but as many as needed. This technique is known as the configuration interaction or superposition of configuration technique; since the first designation is more common, I shall adopt it hereafter (C.I. for short). If we designate the Slater determinants as $\Psi_o, \Psi_1,...,$ the C.I. wavefunction is of the form

$$\Psi = a_o \Psi_o + a_1 \Psi_1 + a_2 \Psi_2 + ... \tag{16}$$

By optimizing in a variety of ways the orbitals in each function Ψ_i and by selecting the C.I. coefficients a_o, a_1, a_2, by applying the Rayleigh-Ritz variational principle

$$\sum_i (H_{ij} - E S_{ij}) a_i = 0 \tag{17}$$

where $H_{ij} = \langle \Psi_i | H | \Psi_j \rangle$ and $S_{ij} = \langle \Psi_i | \Psi_j \rangle$, one obtains a solution Ψ, necessarily as good as, or better than, Ψ_o, and if the series of the above equations is sufficiently long, then we shall reach an exact solution. The only trouble is that the necessary series is very long. The general scheme for configuration interaction calculations was first proposed by Boys in 1950[17] who also carried through a ten configuration calculation on the beryllium atom[18]. Methods similar in philosophy to the CI method had earlier been used by Hartree and Swirles in their study of the oxygen atom[19]. The numerical difficulties, however, prohibited further developments of the theory until the electronic computer came into general use and the problem of integral evaluation and handling of large amounts of numerical data could be efficiently dealt with.

A number of Computer program for ab-initio calculations on molecular systems on the Hartree-Fock level of approximation such as IBMOL and POLYATOMS were constructed in the period 1965-70 and such calculations are today performed routinely in many laboratories. This development has been of great importance for the application of ab-initio quantum mechanical methods in studies of a variety of current problems in molecular chemistry and physics, since the Hartree-Fock method is known to yield reliable results for many chemical processes.

It is customary to differ between two types of correlation effects. The Hartree-Fock model will in certain cases break down due to degeneracies or near degeneracies between several configurations[19]. Such effects frequently occur in calculations of energy surfaces for chemical reactions. They can be dealt with by means of a limited CI expansion which includes all near-degenerate configurations, where the CI expansion coefficients and the orbitals are simultaneously optimized. This procedure constitutes the MC-SCF (Multi-Configurational Self-Consistant-Field) Method[19,20].

The second effect is the dynamic correlation of the electronic motion[21]. For electrons with parallel spins this correlation is already to a large extent included in the Hartree-Fock model, through the anti-symmetry requirement (Fermi hole). The main part of the remaining error is therefore due to correlation of electron pairs having opposite spins. The fact that electron correlation to a good approximation is described by pair interactions implies that the dominant part of the correlation energy is obtained with a CI wavefunction which contains replacements in each pair of electrons. However, single and triple replacements are also important and the wavefunction describing dynamical correlation effects should therefore in general contain the near-degenerate configurations and single and double, and possibly triple, replacements with respect to all of them. Such an expansion will in general be very long since the number of possible double and triple replacement configurations is large for a reasonably sized molecular orbital basis set. One of the obstacles in using the CI method is therefore the problem of calculating and handling a very large number of matrix elements. In special cases it is however possible to avoid the construction of a large Hamiltonian matrix and instead calculate the CI expansion coefficients directly from the given list of molecular one- and two-electron integrals. This is the philosophy behind the CIMI method[22] (direct Configuration Interaction from Molecular Integrals) that has been applied for example to the molecules H_2O, OH^- [23].

1.6) Combination of Statistcal Methods and C.I. Techniques - In the Wigner's proposal [10,13] it is assumed that the correlation correction can be obtained by expanding the exact function in term of a series having the Hartree-Fock function as zero order term.
There are, however, problems in selecting the Hartree-Fock function as reference function. Indeed, despite the fact that the Hartree-Fock model can be uniquely defined, nevertheless the Hartree-Fock function has a variable degree of reliability, when compared with the exact functions (or to the real systems).
An example is provided by the four-electron isoelectronic series, Be, B^{+1}, ... in the $1s^2 2s^2$ configuration: since by increasing the nuclear charge, Z, the Hartree-Fock function becomes more and more hydrogenic and since for a hydrogenic function the 2s orbital is degenerate with the 2p orbital, then the description in the Hartree-Fock model for the four electron problem of Z=4 is superior to the description for much higher Z values[19,20].
From the molecular point of view, there is an additional problem: the Hartree-Fock model often fails to correctly describe the dissociation behavior. This is due, in most cases, to the constraint placed upon the Hartree-Fock model of describing one electron-pair by the same orbital. Thus when the Hartree-Fock function cannot dissociate properly into the atomic products, the function becomes poorer the larger the internuclear separation becomes.
Lie and Clementi[24] have quantitatively elucidated some of the problems encountered in deriving a functional of the density capable of predicting the correlation corrections in molecular systems with a reasonable accuracy and a maximum of simplicity.

An empirical parametrization of the functional of eq. (2) was reported as

$$E_c = \int 0.02096 (1.2 + \rho_m^{1/3})^{-1} \rho_m^{4/3} \, dv + \int 0.02096 \, \ell n (1+2.39 \, \rho_m^{1/3}) \rho_m \, dv \quad (17)$$

where the modified density ρ_m is given by the relation

$$\rho_m = \sum \bar{n}_i \rho_i \quad \text{and} \quad \bar{n}_i = n_i \exp(-0.5(2 - n_i)^2) \quad (18)$$

The main variation of eq. (17) relative to eqs. (13) and (14) is that, in dealing with open shells, the density of an unpaired electron is not taken equal to the computed Hartree-Fock density. As shown, the Hartree-Fock total density, ρ, is the sum of the individual electron's density, ρ_i. In eq. (17) a modified density ρ_m was used rather than the Hartree-Fock density. This modified density is the Hartree-Fock density for a given orbital multiplied by the orbital-occupation number, n_i, (0, 1 or 2 for Hartree-Fock or any value between 0 and 2 in the multi-configuration self-consistent-field) and an exponential expression, $\exp(-0.5(2 - n_i)^2)$, that ensures that $\bar{n}_i$ is equal to 2 when we have a Hartree-Fock pair and a value smaller than 1 for a singly occupied orbital. This stems from the fact that intrapair correlation is, in general, larger than inter-pair correlation. In other words, a singly occupied orbital should not contribute to the total correlation energy as heavily as a paired electron. The functional form of the exponential correction, in principle, can be parametrized as a function of the spin and angular momenta.

The semi-empirical functional of eq. (16) was applied to the second row hydrides LiH($^1\Sigma^+$), BeH($^2\Sigma^+$), BH($^1\Sigma^+$), CH($^2\Pi$), NH($^3\Sigma^-$), OH($^2\Pi$) and HF($^1\Sigma^+$). These molecules with the exception of BeH, dissociate in the Hartree-Fock approximation to incorrect atomic products. Thus, for these molecules the Hartree-Fock is a poor reference function at large internuclear separations. To compensate for this deficiency, Lie and Clementi have added to the Hartree-Fock function as many determinants as needed to obtain the correct dissociation limit, and we shall use this short expansion for all values of the internuclear distances. The technique used is the multi-configuration self-consistent field technique. The resulting function was designated as H-F-P-D; "Hartree-Fock with proper dissociation".

The functional of the density as given in eq. (17) was applied to the H-F-P-D functions. The resulting non-relativistic total energy, i.e., the sum of the correlation energy obtained from eq. (17) and the H-F-

P-D energy are reported in Fig. 2.

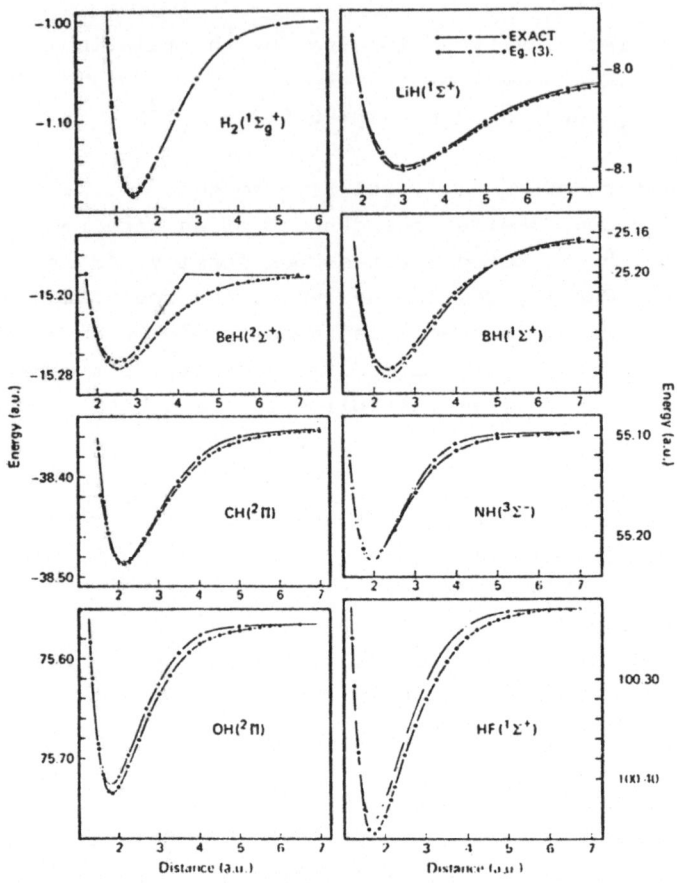

Figure 2 "Exact non-relativistic" and results from Eq. (3) for the hydrides (open circle line). The experimental curves are matched to the calculated values at R = 10 a.u. by subtracting 0.0298, 0.0156, 0.0131. 0.0119, 0.0125, 0.0042 and 0.0057 a.u. from the experimental energies of LiH, BeH, BH, CH, NH, OH and HF, respectively. (For H_2, we shift the calculated value up by 0.0155 a.u. to bring the energy of H_2 at R = 10.00 a.u. to −1.0000 a.u.)

From the data reported in Fig. 2 one can conclude that eq. (17) is a useful relation to predict molecular binding. This statement is confirmed by a second series of computations by the same authors[25] on several diatomic omonuclear molecules.

1.7) <u>Molecular Extra Correlation Energy</u> - From our previous discussion
on the atomic correlation energy it is clear that the correlation energy in a molecule should be related to the correlation energy in the
component atoms. First of all, from the statistical analysis of the
correlation energy we know that the correlation energy is proportional
to the electronic density. From the statistical model we know of the
additivity of the correlation, to a first order. Indeed, the total density is the sum of the orbital densities. Whenever in a molecule an orbital has mainly atomic character (like for the inner shells), its correlation is that of the corresponding atom. Transferrability of correlation energy data from atoms to molecules or solids has been obvious
since 1934[10].

However, there are differences in the density of a molecule when compared with the density one would obtain by simply superimposing atomic
densities. Therefore, we would expect differences in the molecular correlation energies of the corresponding atoms.

The energy difference obtained by subtracting from the correlation
energy of a molecule the correlation energy of the component atoms
(in the correct dissociation state) has been called (Clementi, 1962)
Molecular Extra Correlation Energy[26]. The name implicitly contains
the connotation of an increase in correlation energy from the sum of
the corresponding atom. The main reason of such increase is that the
molecular formation always increases the number of electron pairs.

This increase is two-fold: the most important one is due to the fact
that previously impaired electrons in the separated atoms, pair during
molecular formation. For example, the hydrogen atom has one unpaired
spin, the fluorine atom has 9 electrons, 8 paired and one unpaired,
but in the hydrogen fluoride molecule the two unpaired spin (the one
of H and the one of F) pairs up $\left(H(^2S) + F(^2P) \rightarrow HF\ (^1\Sigma^+)\right)$. Extreme
cases of pairing in diatomic molecules clearly correspond to extreme
unpairing (higher spin multiplicity) of separated atoms like N_2 from
$N(^4S)$, P_2 from $P(^4S)$, Mn_2 from $M_n(^6S)$.

The second factor in increasing the number of pairs is that more inter-
pair correlation energy contributions will occur, because of the increase in the number of electrons.

The molecular extra correlation energy can be a substantial fraction
of the total dissociation energy.

Let us indicate with hf, c, r quantities obtained from the Hartree-Fock
methods, or due to correlation or relativistic effects respectively. Let
us indicate the component atoms of an N atom molecule, m, by the subscript a, and finally let us use the letter D for dissociation energy

and E for total energy.

Then the exact dissociation energy of a molecule is given by the relation

$$D = D_{hf} + D_c + D_r \qquad \text{or} \qquad D = D_{hf} + MECE + D_r \qquad (18)$$

where

$$D_{hf} = \sum_{a=1}^{N} E_{a,hf} - E_{m,hf} \qquad (19)$$

$$D_c = \sum_{a=1}^{N} E_{a,c} - E_{m,c} = \text{Molecular Extra Correlation} \qquad (20)$$
$$\text{Energy}$$

$$D_r = \sum_{a=1}^{N} E_{a,r} - E_{m,r} \qquad (21)$$

In other words, we partition the dissociation energy in Hartree-Fock component, correlation component and relativistic energy component. The quantity D_{hf} is obtained directly from the molecular computations of Hartree-Fock type (now standardly available) and from the atomic Hartree-Fock energies[27,28].

The quantity D_c is the definition of molecular extra correlation energy (MECE) previously discussed.

The quantity D_r is the relativistic part of the dissociation energy. It is customary to assume that $\sum_{a=1}^{N} E_{a,r} = E_{m,r}$ and, therefore, to discount such terms. We do not agree with such an optimistic approach, exception made for molecules containing first and possibly second row atoms. For molecules containing third row atoms, we expect D_r to be from few tenths up to a kilocalorie in extreme cases. However, such contributions can be computed safetly from perturbation theory. For molecules containing fourth row (and higher) atoms, the relativistic contribution to the dissociation energy is not expected to increase nearly as much as the atomic relativistic, however, to increase somewhat more than merely in the 0.1 to few kilocalories range. The above comments applies in particular to compounds where the heavy atom is highly ionic, in the sense given by population analysis.

Let us take a few examples of dissociation energy. For the N_2 molecule in its ground state ($^1\Sigma_g^+$) the experimental dissociation energy is 9.90 e.V. the computed dissociation energy in the Hartree-Fock approximation is D_{hf}=5.18, therefore, the Molecular Extra Correlation Energy is 4.72 e.V. (taking D_r = 0.000 e.V.).

Let us now consider HCN in its ground state ($^1\Sigma^+$). The computed dissociation energy is $D_{hf}=8.85$ e.V., the experimental dissociation energy $D=13.55\pm0.05$ e.V. and MECE=4.7 ± 0.05 e.V. It is noted that the two molecules are isoelectronic and that also for HCN we form three new pairs in the reaction

$$H(^2S) + C(^3P) + N(^4S) \longrightarrow HCN(^1\Sigma^+)$$

The agreement of the extramolecular correlation of 4.72 e.V. for N_2 and 4.70 ± 0.05 e.V. indicates how transferable are data within molecular systems. It is noted that MECE for N_2 is somewhat larger than MECE for HCN : the reason is that in HCN the electronic charges are more localized than in N_2.

1.8) <u>Choice of the Basis Set for SCF-LCAO-MO Functions</u> - Let us briefly comment on the selection criteria for the basis sets. In general, the choice depends on what one wishes to obtain from a wave function. For example, one might wish low accuracy in the total energy, but fair accuracy in the valence electrons, or high accuracy in the inner shells and low accuracy in the valence electrons. Clearly, the strategy in selecting wavefunction accuracy should be a function of the assumed goals. We shall consider the molecule of water as an example.
The basis set for the hydrogen and oxygen atoms are those computed recently by F. van Duijneveldt[29] and are not reported here since too many tables would be required. We have used from 2 to 7 s functions for the hydrogen atom (designated as H/2,3,...7) and for the oxygen atom the smallest basis set is 4/2 namely 4 s and 2 functions of p_x, p_y, p_z type) up to 13/8 (namely 13 s functions and 8 functions of p_x, p_y and p_z type).
With the basis set for atomic hydrogen and oxygen atoms, a number of computations for the water molecule were performed. The molecular geometry for these computations is the experimental equilibrium geometry. In Table 1 the Total Energy (in a.u.) is given for 68 different basis sets; in this table the hydrogen and the oxygen functions have been used without contraction. From the table it is clear that the Hartree-Fock limit for a basis set containing only s and p functions on the oxygen and only s function on the hydrogen is about -76.026 a.u. In addition, if we wish an accuracy of 0.001 a.u. (which is adequate for Hartree-Fock computations) a basis set of 6s functions on the hydrogen and 11s functions plus 7p functions on the oxygen (11s, 7p, 6s) is sufficient. If one is interested in an accuracy of 0.01 a.u., then a set

with 9s, 5p for oxygen and a 5s for hydrogen seems to be adequate.

Table 1

TOTAL ENERGY FOR H_2O (IN a.u.) WITHOUT POLARIZATION*

H \ O	2	3	4	5	6	7
4/2	-75.15053	-75.17413	-75.17890			
5/2	-75.42926	-75.45301	-75.45778			
6/2	-75.50143	-75.52124	-75.52585			
5/3	-75.74707	-75.77026	-75.77419			
6/3	-75.80813	-75.82896	-75.83298			
7/3	-75.86667	-75.88745	-75.89138			
8/3			-75.90618	-75.91008	-75.91171	-75.91212
7/4			-75.97129	-75.97500	-75.97614	-75.97603
8/4			-75.98710	-75.99073	-75.99187	-75.99175
9/4			-75.99126	-75.99489	-75.99603	-75.99592
9/5			-76.01153	-76.01467	-76.01567	-76.01545
10/5			-76.01419	-76.01721	-76.01818	-76.01796
10/6			-76.01824	-76.02118	-76.02199	-76.02178
11/6			-76.01946	-76.02234	-76.02313	-76.02292
12/6			-76.01982	-76.02267	-76.02345	-76.02325
11/7			-76.02088	-76.02381	-76.02457	-76.02435
12/7			-76.02120	-76.02412	-76.02486	-76.02465
13/7				-76.02418	-76.02492	-76.02471
13/8				-76.02499	-76.02573	-76.02552

*The first column X/Y gives the number X of s functions and the number Y of p functions used on the oxygen atom; the numerals 2, 3, ...7 in the first row gives the number of s functions used on each hydrogen atom.

These comments should be taken with care however; there is no good physical reason, really, for selecting an 11s, 7p, 6s set rather than a 12s, 6p, 6s or an 11s, 6p, 6s set. Regarding computational time, however, a 12s, 6s, 6s set is cheaper than an 11s, 7p, 6s set, since in molecule the p functions are used three times ($2p_x$, $2p_y$, $2p_z$). It should be noted that for accurate computations, a larger basis set optimized on the atom does not necessarily guarantee a better S.C.F. energy when used in a molecule, than a somewhat smaller basis set optimized on the atom.

In Table 2 the computed binding energy for H_2O is reported (in a.u.).

Table 2

COMPUTED BINDING ENERGY FOR H_2O (IN a.u.)*

O \ H	2	3	4	5	6	7
4/2	.2037	.2054	.2052			
5/2	.1992	.2010	.2007			
6/2	.1832	.1811	.1807			
5/3	.2225	.2238	.2227			
6/3	.1941	.1930	.1920			
7/3	.1940	.1925	.1917			
8/3			.1915	.1943	.1957	.1960
7/4			.2063	.2090	.2098	.2096
8/4			.2068	.2094	.2102	.2100
9/4			.2066	.2091	.2100	.2098
9/5			.2120	.2141	.2148	.2145
10/5			.2126	.2146	.2152	.2150
10/6			.2131	.2150	.2156	.2153
11/6			.2133	.2151	.2156	.2154
12/6			.2133	.2151	.2156	.2153
11/7			.2138	.2157	.2162	.2159
12/7			.2138	.2157	.2161	.2158
13/7				.2156	.2161	.2158
13/8				.2162	.2166	.2164

*These values are obtained by subtracting from the total energy given in Table 10, the atomic oxygen total energy (^{3}p) and twice the hydrogen total energy (^{2}S) as given in reference (5c).

The reported values are the difference between the Total Energy of H_2O (Table 1) and the sum of the atomic energies of the hydrogen and oxygen atom, computed with the same basis as in the molecule. The resulting binding energy is nearly constant for all basis sets. Thus from this example, it follows that the computed binding energy is nearly constant for very different basis sets as long as (1) the binding energy is defined as above, (2) no polarization functions are introduced, and (3) an equally balanced set of functions for all atoms in the molecules is selected. Equally balanced set means: poor atomic set for all atoms, medium set for all atoms, good atomic set for all the atoms in the molecule.

Table 3

POPULATION ANALYSIS AND DIPOLE MOMENT (IN a.u.) FOR H_2O WITHOUT POLARIZATION

Case	H charge	O charge	Dipole moment
0:4s,2p;H:4s	0.7824	8.4352	0.7965
0:5s,2p;H:4s	0.7794	8.4411	0.7994
0:6s,2p;H:4s	0.7537	8.4925	0.7833
0:5s,3p;H:4s	0.6906	8.6188	0.9757
0:6s,3p;H:4s	0.6446	8.7107	0.9628
0:7s,3p;H:4s	0.6427	8.7145	0.9623
0:8s,3p;H:4s	0.6426	8.7147	0.9617
0:7s,4p;H:4s	0.6275	8.7450	1.0123
0:8s,4p;H:4s	0.6271	8.7458	1.0138
0:9s,4p;H:4s	0.6275	8.7450	1.0129
0:9s,5p;H:4s	0.6251	8.7498	1.0408
0:10s,5p;H:4s	0.6339	8.7221	1.0399
0:10s,6p;H:4s	0.6293	8.7494	1.0526
0:11s,6p;H:4s	0.6351	8.7298	1.0531
0:12s,6p;H:4s	0.6377	8.7246	1.0532
0:11s,7p;H:4s	0.6303	8.7394	1.0590
0:12s,7p;H:4s	0.6320	8.7359	1.0592
0:12s,7p;H:5s	0.6239	8.7522	1.0609
0:12s,7p;H:6s	0.6180	8.7640	1.0611
0:12s,7p;H:7s	0.6185	8.7629	1.0618

In Table 3 we continue our analysis for the wave functions reported in Table 1. This time the gross atomic charges (from the electronic analysis) and the dipole moment (in a.u.) are reported for a subset of the cases given in Table 1. One can notice that with exception of the cases where only two 2p Gaussians were used for the oxygen atom, the dipole moment oscillates between 0.96 a.u. and 1.06 a.u. This variation of 10% is systematic and with an increased set, one quickly approaches the limit value of 1.06 a.u. The gross charges converge to a value of 0.63 on the hydrogen atom and 8.74 on the oxygen atom. Again, the convergency is regular and fast.

Thus, in selecting a basis set, if one whishes to predict dipole moments and charges, a rather large set must be used in order to converge to a limit.

We now introduce polarization functions. A $2p_x$, $2p_y$, $2p_z$ basis with orbital exponent 0.75 was added to the hydrogen atoms, and a $3d_{xx}$, $3d_{yx}$, $3d_{yy}$, $3d_{zx}$, $3d_{zy}$, $3d_{zz}$ was added to the oxygen atom (with orbital exponent 1.0).

Table 4

EFFECT OF POLARIZATION FUNCTIONS ON H_2O TOTAL ENERGY.*

Case	sp basis	sp + polarization	Difference
0:4s,2p;H:4s	-75.17890	-75.29354	.11464
0:5s,2p;H:4s	-75.45778	-75.54615	.08837
0:6s,2p;H:4s	-75.52585	-75.57691	.05106
0:5s,3p;H:4s	-75.77419	-75.85941	.08522
0:6s,3p;H:4s	-75.83298	-75.88177	.04879
0:7s,3p;H:4s	-75.89138	-75.93602	.04464
0:8s,3p;H:4s	-75.90618	-75.94976	.04358
0:7s,4p;H:4s	-75.97129	-76.01274	.04145
0:8s,4p;H:4s	-75.98710	-76.02657	.03947
0:9s,4p;H:4s	-75.99126	-76.03032	.03906
0:9s,5p;H:4s	-76.01153	-76.04867	.03714
0:10s,5p;H:4s	-76.01419	-76.05047	.03628
0:10s,6p;H:4s	-76.01824	-76.05498	.03674
0:11s,6p;H:4s	-76.01946	-76.05597	.03651
0:12s,6p;H:4s	-76.01982	-76.05630	.03648
0:11s,7p;H:4s	-76.02088	-76.05778	.03690
0:12s,7p;H:4s	-76.02120	-76.05811	.03691
0:12s,7p;H:5s	-76.02412	-76.05952	.03540
0:12s,7p;H:6s	-76.02486	-76.05973	.03487
0:12s,7p;H:7s	-76.02405	-76.05985	.03580

*The added polarization functions are $2p_x$, $2p_y$, $2p_z$ on the hydrogen atoms (with exponent 0.75) and $3d_{xx}$, $3d_{yx}$, $3d_{xy}$, $3d_{zx}$, $3d_{zy}$, $3d_{zz}$ on the oxygen (with exponent 1.0).

In Table 4, the gain in total energy is reported.

In Table 5, we report gross atomic charges and the dipole moment for the new basis set (with polarization functions).

Table 5

POPULATION ANALYSIS AND DIPOLE MOMENT FOR H_2O (BASIS SET WITH POLARIZATION)

Case	H charge	O charge	Dipole moment (a.u.)
0:4s,2p;H:4s	0.8154	8.3692	0.5688
0:5s,2p;H:4s	0.8160	8.3681	0.5636
0:6s,2p;H:4s	0.8061	8.3878	0.5963
0:5s,3p;H:4s	0.8153	8.3695	0.7615
0:6s,3p;H:4s	0.8065	8.3870	0.7691
0:7s,3p;H:4s	0.8025	8.3950	0.7716
0:8s,3p;H:4s	0.8026	8.3948	0.7714
0:7s,4p;H:4s	0.7490	8.5021	0.8286
0:8s,4p;H:4s	0.7434	8.5132	0.8310
0:9s,4p;H:4s	0.7442	8.5116	0.8301
0:9s,5p;H:4s	0.7098	8.5804	0.8570
0:10s,5p;H:4s	0.7046	8.5909	0.8561
0:10s,6p;H:4s	0.7080	8.5840	0.8662
0:11s,6p;H:4s	0.7037	8.5926	0.8664
0:12s,6p;H:4s	0.7073	8.5854	0.8706
0:11s,7p;H:4s	0.7038	8.5923	0.8664
0:12s,7p;H:4s	0.7067	8.5866	0.8707
0:12s,7p;H:5s	0.7082	8.5836	0.8694
0:12s,7p;H:6s	0.7093	8.5814	0.8693
0:12s,7p;H:7s	0.7066	8.5868	0.8694

From Table 4 we learn that the introduction of polarization functions brings about a gain in the binding energy of about 0.035 a.u. (or nearly 20% of the binding computed in Table 2). One can notice that the difference in total energy between the set with and without polarization decreases from 0.114 a.u. to O.035 a.u. This is a clear indication that the polarization functions added to a small basis set, even if balanced, operate in two complementary ways. On one hand, the polarization functions in-

troduce polarization effects. On the other hand, the polarization functions supplement part of the s and p basis set which is lacking in the
small set. (It is noted that a combination of functions like

$$(3d_{xx} + 3d_{yy} + 3d_{zz})$$

are equivalent to an s function). From Table 4 one can easily conclude
that molecular computations at the Hartree-Fock level with basis sets
which include polarization functions but have insufficient atomic sets
are not too meaningful if used to predict the Hartree-Fock binding energy. The dipole moment and atomic gross charges (Table 5) present a rather systematic trend. If one wishes to compute a dipole moment within
0.02 a.u. (or 0.05 Debye units) from the Hartree-Fock limit, then one
should use at least a 9s, 5p set for oxygen, and 4s for hydrogen plus
polarization functions. The gross charges do not fluctuate as long as
one uses a basis set both balanced and larger or equal to a 9s, 5p on
oxyge, and 4s on hydrogen. It is noted that the Hartree-Fock model predicts dipole moments accurate within 0.25 or 0.4 debyes. Computations
of dipole moments reaching an agreement with experimental data better
than the above limit are often either fortuitous or are artifacts obtained by conveniently varying the orbital exponent of the polarization functions, so as to obtain agreement with experimental data.

Table 6

TOTAL ENERGY FOR H_2O WITH POLARIZATION*

Polarization Function Exponents				Energy (a.u.)	Energy Gain (a.u.)
Od	Of	Hp	Hd		
				-76.02563	0.0
1.00		0.75		-76.06027	0.03464
0.40 2.00		0.75		-76.06186	0.03623
0.40 2.00		0.30 1.50		-76.06400	0.03837
0.40 1.00 2.00		0.30 1.50		-76.06454	0.03891
0.40 2.00	1.00	0.30 1.50		-76.06503	0.03940
0.40 1.00 2.00	1.00	0.30 1.50	0.75	-76.06587	0.04024

In Table 6 H_2O energies (in a.u.) computed with large basis sets are

reported. We started with an (9 13s, 8p; H 6s) basis set and found
that contracting the six O s functions, the four O p functions and the
three H s functions with the largest exponents resulted in an energy
loss of only 0.00010 a.u. compared with the energy of −76.02573 a.u.
in Table 1 for the completely uncontracted basis set. Consequently we
used a (O 8s, 5p; H 4s) contracted basis set and augmented it with
the polarization functions given in Table 6. The best energy   of
−76.06587 a.u. was obtained by using three sets of d functions and a
set of f functions on the oxygen and two sets of p functions and a set
of d functions on the hydrogens. A consideration of the energies reported in Tables 4 and 6 leads us to believe that the above energy is
about 0.002 a.u. from the Hartree-Fock limit for H_2O with the experimental equilibrium geometry.

1.9) <u>Electron Population Analysis</u> − The SCF wavefunctions can be analyzed indirectly via a study of the physical properties of the molecule under consideration (like moments, polarizabilities, vibrational
analysis, etc.) or directly by what is known as "electron population
analysis". In the following, we shall briefly expose the method of
Mulliken[30] here somewhat modified and extended.
From the previous exposition of the SCF approximation, a molecular orbital is written as

$$\varphi(\lambda i) = \sum_p c'(\lambda ip)\, \chi'(\lambda p) \qquad (22)$$

where λ, i, p are indices which refer to symmetry representation, a
specific orbital, and a specific basis set, respectively. The basis
set is in general a symmetry adapted function (SAF), i.e., it trans-
forms as λ. In the LCAO approximation the $\chi'(\lambda p)$ is a linear combination of functions, designated by χ_q, centered on the atoms. The linear combination coefficients of the SAF are determined on the basis
of symmetry alone, and we can write

$$\varphi(\lambda i) = \sum_p c''(\lambda ip) \sum_q d(\lambda pq)\chi(q) \qquad (23)$$

By combining the c'' and the d into a new coefficient c we have

$$\varphi(\lambda i) = \sum_{ms} c(\lambda ims)\chi(ms) \qquad (24)$$

where for each λ and i the index m refers to a given atom and the index
s refers to a given X_q on the m atom.

For real functions, the electronic density of $\varphi(\lambda i)$ is

$$\langle \varphi(\lambda i)/\varphi(\lambda i)\rangle = \sum_{ms} \sum_{m's'} c(\lambda ims)c(\lambda im's')\langle \chi(ms)\chi(m's')\rangle \qquad (25)$$

This relation is the base of Mulliken's analysis. The above sum can be written as

$$\langle \varphi(\lambda_i)|\varphi(\lambda_i)\rangle = \sum_{mm'} \sum_{ss'} (mm'|ss')_{\lambda i} \qquad (26)$$

$$\text{or} \qquad = \sum_{mm'} (mm')_i \qquad (27)$$

when no symmetry is taken into account, and where in general for m=m' and s=s' the overlap $\langle \chi_s|\chi_{s'}\rangle$ is unity since the SAF as well as the φ are normalized. For each MO, a table can be constructed. There is one such table for each MO, $\varphi_{\lambda i}$, and each table is by construction symmetrical. We shall call "quadrant" the matrix of numbers with a given m and m' and indicate this as $(mm')_{\lambda i}$. The sum of its terms is indicated as $S(mm')_{\lambda i}$. The diagonal elements of a quadrant are indicated as $(mm')_{d\lambda i}$ and its sum as $S(mm')_{d\,i}$.
For m=m', the quadrant $(mm)_{\lambda i}$ contains quantities which are specific to the atomic set fpr the atom m; for m≠m' the quadrant $(mm')_{\lambda i}$ contains quantities which are specific to atomic sets of the atoms m and m'.
For the atom m the following definitions, borrowed from R.S. Mulliken, are given [30]:

Net atomic population
$$P_m = \sum_{\lambda} \sum_{i} S(mm)_{\lambda i} \qquad (28)$$

Overlap population with atom m'
$$P_{mm'} = \sum_{\lambda} \sum_{i} S(mm')_{\lambda i} \qquad (29)$$

Gross atomic population
$$G_m = P_m + \sum_{m'} P_{mm'} \qquad (30)$$

Let us now focus our attention on the quadrant (mm). The atomic set (χ's) of any such quadrant will be of s, p, d, etc. type; therefore, within the (mm) quadrant, we can have subquadrants of the type $(m_s m_s)$, $(m_s m_p)$, $(m_p m_p)$, etc. designated in general as $(m_l m_{l'})$ where l and l' are the angular quantum numbers for the χ's. In full analogy to the previous definitions for the quadrants (mm') we can define $S(m_l m_{l'})_{d\lambda i}$

for the subquadrants $(m_1 m_1')$. With this in mind we can introduce the following definitions:

Non-hybrid net atomic $\qquad P_{ml} = \sum_{\lambda} \sum_{i} S(m_1 m_1)_{\lambda i}$ (31)

Hybrid net atomic $\qquad P_{ml\ 1'} = \sum_{\lambda} \sum_{i} S(m_1 m_1')_{\lambda i}$ (32)

Non-hybrid overlap $\qquad P_{ml\ m'l} = \sum_{\lambda} \sum_{i} S(m_1 m_1')_{\lambda i}$ (33)

Hybrid overlap $\qquad P_{ml\ m'l} = \sum_{i} S(m_1 m_1')_{\lambda i}$ (34)

Non-hybrid gross atomic $G_{ml} = P_{ml} + \sum_{\lambda} \sum_{i} \sum_{m'} P_{ml\ m'l}$ (35)

Hybrid gross atomic $\qquad G_{ml\ 1'} = P_{ml\ 1'} + \sum_{\lambda} \sum_{i} \sum_{m'} P_{ml\ m'l'}$ (36)

Hybridization is a very familiar concept in theoretical chemistry. However, its meaning is often used in an exceedingly restrictive sense, usually when we have more than one atom. However, hybridization includes polarization, and therefore we can talk of hybridization between two atoms or between two electrons on the same atom. As a consequence we have internal hybridization (within a given atom and due to the electrons of that atom) as well as external hybridization (within a given atom and due to a field originated outside the atom). External hybridization is the familiar one. An example of internal hybridization is given by the beryllium ground state atom; the 2s orbital is strongly hybridized (internally) withthe 2p orbitals. Therefore, the correlation problem in atoms can be viewed as a problem of describing in the best possible way the internal hybridization, and the correlation problem in molecules can be viewed as a problem of describing in the best possible way the internal polarization of the component atoms plus the external hybridization.

It is stressed that the above definitions have meaning only for a given basis set. Therefore, they provide quantitative data of qualitative character. However, it is exactly this type of data which we like to analyze in order to obtain some correlation between molecular structures. An exact wavefunction for a molecule provides a tool for obtaining exact expectation values. These can be obtained, as an alternative, from experimental data. However, taken alone, neither an exact list of expectation values nor an accurate list of experimental data constitutes understanding of the electronic structure of molecules.

In Tables 3 and 5 we have presented a systematic study of the variation of the net atomic charges on the hydrogen and on the oxygen atoms
in the water molecule using different basis sets. One can see by inspection how much the basis set effects the computed net atomic charges.
We can summarize this section by stating that Mulliken's electron population analyses represent an efficient way to condense information
contained in the wavefunction. Perhaps it is even more important to
recall that the electron population analysis provides a simple link
between the description of a chemical system as a set of m nuclei and
n electrons and the traditional description of a chemical system as a
set of m atoms.

In the following we shall indicate that the population analyses can be
used to estimate a number of quantities that cannot be otherwise obtained, at present. Let us consider the dipole moment of a molecule.
From long time[31] we know that the dipole moment can be estimated by
using the computed net charges. The limits of the accuracy in this
estimate can be extracted by the data presented in Table 7, where we
compare, for the naturally occurring amino acids, the dipole moments
exactly computed by operating on the way functions[32] and the dipole
moment estimated by using the atomic net charges (computed from the
same wavefunctions). The overall agreement for the data in Table 7 on
the total dipole moment computed in the two ways is within about 20 or
40%.

TABLE 7

Dipole moment (in Debyes) for selected amino acids obtained from a quantum mechanical computation (Q.M.) or estimated from the net charges (N.C.).

Names		Q.M.	N.C.	Names		Q.M.	N.C.
PHE	F	1.49	1.35	GLY	G	1.95	1.69
VAL	V	2.27	2.48	LEU	L	2.50	2.65
ILE	I	0.93	1.25	TRY	O	2.46	2.16
TYR	W	-	1.90	THR	T	2.69	2.50
SER	S	2.64	3.71	ARG	R	_	5.01
ASN	Q	5.68	6.22	LYS	K	2.77	3.34
GLN	N	3.18	2.03	GLU	E	2.70	2.87
ASP	D	2.57	1.99	PRO	P	2.18	2.78
HPR	-	2.44	2.78	CYS	C	5.21	4.87
CSH	-	3.46	2.69	MET	M	1.54	1.04
ALA	A	1.78	1.37	HYS	H	-	4.91

Let us now consider a complex chemical system, that today can not be described by a wavefunction of the SCF-LCAO-MO type; as example let us consider the enzyme lysozyme. If we wish to estimate its dipole moment we can make use of the atomic net charges computed for the separated amino acids and for some reasonable system, like three glicines[33], to simulate an element of the back-bone atoms connecting the amino acids in protein. In this way we have obtained a dipole moment of about 8 Debyes for the enzyme lysozyme, making use of the crystallographic data provided us by A. Yonath[34].

Another example of the use of atomic net charges for estimating quantities depending on the electronic structure is provided by the computation of the electrical potential V and of the electric field E at any point within an enzyme (and of an enzyme with its substrate). This problem is of importance because the reaction between an enzyme and its substrate is an event constrained not only by the geometry of the two interacting species, but also by the possible existence of resulting field at the active site, originated by the residual charges on each atom of the enzyme and of its substrate[35].

We note that it is well known that an enzyme activity drops sharply outside a given pH range. For a given value of the pH, a number of amino acids are ionized and tend to regain electroneutrality either by forming internal salts among themselves or by attracting a counter ion (always present in physiological water and even in the enzyme in crystalline form) and by properly orienting the molecules of water around the charged region. Thus the active enzyme will have an electric field at any given position due not only to the residual charges of the neutral form of the enzyme and its substrate, but also of the partially neutralized charges of the ionized amino acid and its counter-ion. The effect of the electric field intensity on the enzymatic activity constitutes an open question, that can be partially answered by a knowledge of atomic net charges.

In Fig. 3 we report a cross section of the enzyme lysozyme that cuts through the reactive site. In Fig. 3 we report the isopotential curves and the "borders" of the enzyme. The enzyme structure is the one in the triclinic crystalline form (see reference 34); a number of parallel planes of which Fig. 3 represent one example, are selected to cut across the enzyme. A computer program subdivides the plane with a square grid of points and at each point a search is made to determine, if within 1 A of the grid point there is an atom of the enzyme. The "borders" of the enzyme are built by connecting those grid points that

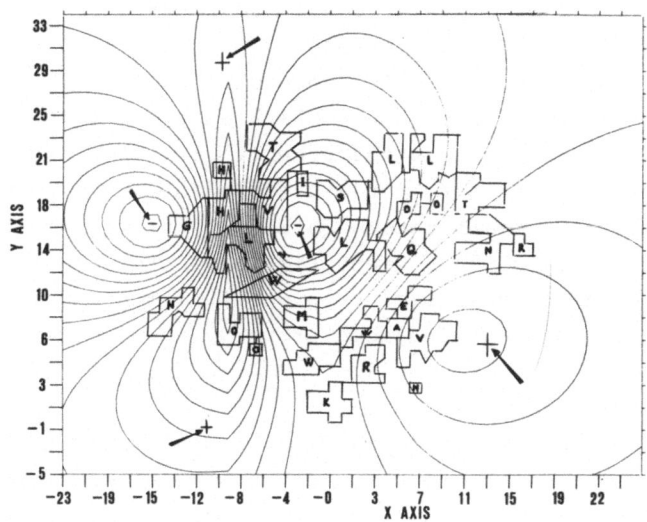

Fig. 3 - Isopotential contours for a plane bisecting the lysozyme enzyme in the vicinity of the reactive site. The areas of irregular shape containing an alphabetic character are amino acid on the selected plane or within 1 Å from the selected plane (see text and Table 7 for additional explanations).

i) have an atom within a sphere of 1 A radius centered at the grid point and ii) are immediately adjacent to a grid point where there is no atom of the enzyme within a sphere of radius 1 A, centered at the grid point. For each grid point we have computed the quantity $V(r) = \sum_i q_i/r$, where r is the distance of the grid point from all the atoms of the lysozime molecule having atomic net charge q. In addition for each grid point we have computed the electric field intensity, E, and its components $E(x)$, $E(y)$ and $E(z)$. The field expressed either in terms of $V(r)$ or in terms of E (or its component) is clearly strong in the immediate vicinity of those atoms that are very near or on the selected plane; for this reason the field and the potential are not

computed within a cut off sphere, centered around each atom; as a consequence in Fig. 3, the iso-potentials are not given for the values
above a given threshold (these iso-countours would be of no interest,
and would only appear in Fig. 3 as a black spot resulting from the
many iso-contours one very near the other). From Fig. 3 one can see
that the potential is non zero for a sizable region of the cross section
through the enzyme. We shall return to this problem in the last chapter of this work; here we have been primarily interested in presenting
the electron population analysis for a wavefunction and to indicate
how one can use it to determine some property in very complex molecular systems[35].

1.10) <u>Bond Energy Analyses in a Single Molecule</u> - The set of m atoms constituting a molecule can be represented in several ways; the simplest
representation is limited to a statement of the number and types (Z
value) of atoms constituting the molecule namely the gross chemical formula. A more refined representation schematically adds to the gross
formula the geometrical relationship for the elements of the set
(structural formula); in this representation among the $m(m-1)/2$ con-
nections, representing pairwise interactions, only few are selected
and precisely those among nearest neighboors and in such number as to
satisfy the valence of the m atoms. In this way the "connection" can
be either single or multiple (single bonds, or multiple bonds).
The organization of the atoms into the Atomic Periodic Table (1869)
provided the first theoretical base for a representation of a molecule and with Van't Hoff findings (1874) the bases were set for the
representation of a molecule as a sterical structural formula. The
next important step is provided by G.N. Lewis "electronic rationali-
zation" of a chemical compound with the introduction of the concept
of an electron pair, placed between two bonded atoms, and with a pre-
quantum-mechanical explanation of ionic and covalent bonds. Finally,
by formulating the concept of a "lone pair", G. N. Lewis introduced
the concept that today we could define of "actual" and "latent" bonds
for a chemical reaction. A quantum mechanical description of the electronic structure of a molecule followed soon with the LCAO-MO concepts
by R.S. Mulliken (1935), explicitly recognized as an approximation to
the Hartree-Fock representation (taken as a first step towards more
accurate wavefunction). Much later Mulliken attempted to reconcile
with the population analysis formalism (1955) the traditional idea
that a molecule is composed by a set of m atoms; in the MO basic assumption a molecule is composed by a set of m fixed nuclei and n elec-

trons.

But it was not possible to obtain from the wavefunction a representation equivalent to the one provided by the structural formulas. The chemist needs to associate with chemical bond two concepts, namely, the electronic density and a bond energy. The statement "forming", "breaking" a bond implies both density and energy and in rather a subtle way since the bond energy can be either attractive or repulsive, whereas the density can only be positive. For this reason we have introduced the bond energy analysis approximation (1967). In this work the early concepts are summarized and somewhat extended. We start by recalling that the molecular orbital $\varphi(i\lambda)$ (or simply $\varphi(i)$, for simplicity in writing) can be written as a linear combination of a set of functions that are centered on the nuclei (or, if not centered on the nuclei can always be expanded into an equivalent set with the origin at the nuclei). The bare nuclei hamiltonian matrix elements (see eq. (3)) $\langle \varphi(i)| h_o|\varphi(i)\rangle$ are therefore expressed as matrix elements over the basis set characterized by at most three nuclear positions, two (or one) related to the basis set origin, one (or none) related to the hamiltonian. The two electron operators of the Hamiltonian (namely, $1/r(1,2)$) will equivalently be characterized by matrix elements over the basis set with at most four nuclear positions (we shall use the word "center" as a short expression for the equivalent one "nuclear position where a basis set is centered"). If we designate the m nuclei with indices like A or B or C or D, then the total energy of the moelcule, usually expressed as in eq. (4) can be rewritten as follows:

$$E = \sum_A E(A) + \sum_{A\neq B} E(A,B) + \sum_{A\neq B\neq C} E(A,B,C) + \sum_{A\neq B\neq C\neq D} E(A,B,C,D)$$

where the last summation drops out for three atomic molecules and the last two summations drop out for diatomic molecules. Thus, the total energy for any molecule can be written as the sum of one-center energies, two-center energies, three-center energies and four-center energies (with obvious notation)

$$E = E_1 + E_2 + E_3 + E_4$$

The terms in E_1 constitute the sum of the atomic energies of the atoms when in the molecule. Clearly, each term $E(A)$, $E(B)$,... has a higher energy than the corresponding atom in the ground state. The terms in E_2 contain both the nuclear-nuclear repulsion (positive energy term) as

well all those matrix elements where two nuclei appear as indices.
The terms $E(A,B)$, $E(A,C)$, $E(B,C)$, ... of E_2 are in general attractive
(negative sign). The three and four center terms are in general much
smaller than the one and two center terms; the four center terms are
due only to the electron-electron repulsions.

Let us consider in more detail the one-center bond energy term. Since
the early thirties, there have been attempts to describe some aspect
of the electronic structure of molecules, by making use of the concept
of valency state [36,37,38]. The concept of valency state is most
naturally obtained in the Valency Bond approximation[36,38] but can
also be defined in the Molecular Orbital approximation[37,36]. The
valency state concept postulates a fictitious atomic state (with no re-
gard to maintaining L and S as good quantum numbers) which reproduces
as closely as possible the electronic distribution of an atom when in
a molecule. Thus, the valency state concept is, in its origin, connected with the assumption that there "are atoms in molecules", or at
least, that we can identify atomic substructures in a molecular electronic structure.

In this work, we distinguish between the valency state of an atom at
infinite separation from the remaining atoms of a molecule, and the valency state of an atom at finite (i.e., near equilibrium) separation
from the remaining atoms of a molecule. We refer to the former type as
Valency State Standard (V.S.S.), to the latter as Molecular Orbital
Valency State (M.O.V.S.).

More exactly, we define the V.S.S. as the one which can be obtained
in the Hartree-Fock formalism, by relaxing the constraint that L and
S are good quantum numbers and by making appropriate mixtures of
pure states.

We defined the energy of the M.O.V.S. as the one-center energy obtained
from the bond energy analysis. The M.O.V.S. for a given atom will
vary in energy from molecule to molecule, depending on the neighbors
and its molecular geometry. The reason for this choice of the definition of M.O.V.S. is rather simple: if we wish to decompose the total
energy of a molecule into atomic components and atom to atom interactions, than we might as well use a formalism which is simple in its
understanding and directly obtained from a molecular computation. We
realize however that our definition is an arbitrary one, since any
decomposition in partial energies for the total energy of a system
of interacting particles is arbitrary. Before proceeding it might be
useful to introduce few more definitions: by denoting with ε_i the

orbital energy of $\varphi(i)$ we have the following decomposition:

$$\mathcal{E}_i = \sum_A \mathcal{E}_i(A) + \sum_A \sum_B \mathcal{E}_i(A,B) + \sum_A \sum_B \sum_C \mathcal{E}_i(A,B,C) + \sum_A{}' \sum_B{}' \sum_C{}' \sum_D \mathcal{E}_i(A,B,C,D) \tag{37}$$

$$\langle h_i \varphi_i | -\tfrac{1}{2}\bar{\nabla} + \tfrac{Z}{R} | \varphi_i \rangle = \sum_A h_i(A) + \sum_A \sum_B {}' h_i(A,B) + \sum_A \sum_B {}' \sum_C {}' h_i(A,B,C) \tag{38}$$

$$E = \sum_i \sum_A \eta_i(A) + \sum_A \sum_B \eta_i(A,B) + \sum_A \sum_B{}' \sum_C{}' \eta_i(A,B,C) + \sum_A \sum_B{}' \sum_C{}' \sum_D{}' \eta_i(A,B,C,D) \tag{39}$$

where

$$\eta_i(A) = \mathcal{E}_i(A) + h_i(A)$$

$$\eta_i(A,B) = \mathcal{E}_i(A,B) + h_i(A,B)$$

$$\eta_i(A,B,C) = \mathcal{E}_i(A,B,C) + h_i(A,B,C)$$

$$\eta_i(A,B,C,D) = \mathcal{E}_i(A,B,C,D)$$

Whereas the sum of the "orbital energies", $\mathcal{E}$, is not equal to the total energy E, the sum of the "electron's energies", η , is equal to the total energy. It is noted that eq. (39) holds for closed shells and requires some rearrangement for open shell structures. The M.O.V.S. for an atom A is defined by the quantity:

$$E_A = \sum_i \eta_i(A) = \sum_i (\mathcal{E}_i(A) + h_i(A)) \tag{40}$$

Let us consider water as an example. Our best Hartree-Fock computed energy is -76.06587 a.u. [39]; this value was obtained by incrementing a (0:13s, 8p; H:6s) set with polarization functions on the hydrogens (two sets of p functions and a set of d functions) and on the oxygen (three sets of d functions and a set of f cuntions; see Table 6).
The 10 electrons of H_2O, grouped in five molecular orbitals have the orbital energies given in Table 8; in this table the orbital energies are decomposed into 1-, 2- and 3-center contributions according to the B.E.A. (bond energy analyses) prescription. The notation H(1)-O indicates a 2-center contribution between H(1) and O, equivalently H(1)-H(2)-O designate a 3-center contribution. Table 8 clearly indicates that the first orbital is mainly due to the oxygen atom with non-negligible influence from the 2-center contribution along the OH bond

TABLE 8

Orbital Energy Decomposition for H_2O (in a.u.)

Orbital Energy	Decomposition				
	H(1)	O	H(1)-H(2)	H(1)-O	H(1)-H(2)-O
ε_1 = -20.5604	0.0	-20.8171	0.0	0.1502	-0.0437
ε_2 = -1.3511	-0.0033	-0.8431	-0.0211	-0.2392	-0.0017
ε_3 = -0.7179	-0.0374	-0.1426	0.0019	-0.3245	0.1465
ε_4 = -0.5811	-0.0110	-0.6091	-0.0428	0.0652	-0.0376
ε_5 = -0.5082	0.0010	-0.5837	-0.0005	0.0528	-0.0316

and a smaller 3-center contribution. The zero contribution of H(1)
(and H(2)) indicates that there is no electronic charge at the hydrogens for this orbital. The above data are all in good agreement
with our picture that the first orbital is essentially a $1s^2$ electron pair centered at the oxygen nucleus. It is noted that the orbital energy for the $1s^2$ electrons of the oxygen atom in the ground
state configuration is -20.6686 a.u., -20.6932 a.u., and -20.7304 a.u.
for the ^3P, ^1D and ^1S states respectively[28].
As known, the second orbital is mainly a $2s^2$ on the oxygen, whereas
the third and fourth are mainly the bonding orbitals between O and H,
and the fifth orbital is mainly a lone pair electron of the 2p type,
perpendicular to the molecular plane. The orbital energy of the 2s
electrons in the oxygen atom is -1.2443 a.u., -1.2565 a.u. and -1.2751
a.u. for the ^3P, ^1D and ^1S states, respectively. Table 8 indicates
that to equate the orbital energy of separated atoms with the orbital
energy of a molecule can be misleading. The orbital energy of the second orbital in H_2O is -1.3511 a.u., but its decomposition reveals
that this near equality to the value of 1.25 a.u. for the separated
atom is the result of a rather drastic reorganization (with respect
to the separated atom), whereby the O-H term is significantly important.
The two bonding orbitals (third and fourth) are near in energy but

very different in their decomposition. The dominant terms of the third
orbital are the two 2-center H-O terms, whereas the dominant term in
the fourth orbital is associated with the one-center energy (on the
oxygen atom (2p)).
From Table 8 the molecule of water appears to be a somewhat deformed
O^{--} ion with two protons. The five orbitals, $\varphi_1 = \sim 1s^2(O)$, $\varphi_2 = \sim 2s^2(O)$,
$\varphi_3 = \sim 2p_y^2(O)$ and $\varphi_5 = \sim 2p_z^2(O)$. This crude picture does not account
for the non-linear structure of water. Before commenting on this point,
we continue with the analysis of the water molecule (at the equilibrium geometry). The total energy of the molecule is not the sum of
the orbital energies, but the sum of the orbital energies, ε's, and of
the one electron energies, h's, (kinetic+nuclear electron interaction):
$E = \sum(\varepsilon_i + h_i) = \sum \eta_i$ where $\eta_i = \varepsilon_i + h_i$. In Table 9, we report the break-
down of the η's in terms of 1-, 2- and 3- center components.

TABLE 9

Electron Energy Decomposition for H_2O (in a.u.)

η	Decomposition				
	H(1)	O	H(1)-H(2)	H(1)-O	H(1)-H(2)-O
$\eta_1 = -53.6088$	0.0	-52.7277	0.0	-0.4086	-0.0439
$\eta_2 = -9.2098$	-0.0111	-0.89246	-0.0501	-0.3754	-0.3241
$\eta_3 = -7.4404$	-0.0959	-3.1344	-0.0077	-2.2463	0.3859
$\eta_4 = -7.4692$	-0.2917	-6.0643	-0.1016	-0.5376	-0.1798
$\eta_5 = -7.5441$	+0.0018	-6.4608	-0.0012	-0.6234	-0.0467

The first point we notice in Table 9 is that the order of the three or-
bitals φ_3, φ_5 is inverted. φ_5 is the most stable, φ_4 the second stable
and φ_3 the less stable. If we ask what the energy of an electron is in
the molecule, then Table 9 is more informative than Table 8, (despite
a long tradition accustomed to analyzing the equivalent of Table 8 and
ignore the existence of the equivalent of Table 9). This reversal is
not surprising: our rough model which equates H_2O with O^{--}, plus two
protons, has to make use of two electrons of the two hydrogens in or-

der to construct an additional 2p orbital: thus ψ_3 is the "most distorted 2p orbital" of the three pairs of orbitals. On the other hand, ψ_5 is the least distorted. Therefore, since "distortions" cost energy, ψ_5 is the most stable, ψ_3 the least stable, and ψ_4 is intermediate between the two.

The second point to notice is that the last three orbitals are nearly equivalent in binding energy (about 1% variation between ψ_3, ψ_4 and ψ_5), consistent with the "O^{--} plus two protons" picture. In Table 8 ψ_3 is mainly formed by the O-H 2-center terms (-0.6490 a.u. compared to $\eta_3 = 0.7179$); in Table 9, ψ_3 is the result of both the 1-center term at the O site, and the two 2-center terms of O-H type. The main repulsive term in both Table 1 and Table 2 is the 3-center term H-O-H in ψ_3. Tentatively, we shall postulate that whereas the 2-center terms are of predominant importance in defining bond length, 3-center terms are predominantly important for bond angles. Tentatively, we explain the known geometry of H_2O, as the best compromise to reduce the repulsion of the 3-center term[40].

The bond formation diagram for the H_2O molecule in the Hartree-Fock approximation is given in Fig. 4 . As the origin we take the sum of the Hartree-Fock energies of oxygen in the ^3P state (-74.8093 a.u.) and hydrogen in the ^2S state (-0.5000 a.u.). The 1-center energies (for H_2O) summed over the five orbitals are -74.1525 a.u. and -0.1342 a.u. for the oxygen and for each hydrogen atom, respectively. Thus, the 1-center sum is 1.3884 a.u. above the previously defined zero. The total 2-center energies are 0.1880 a.u. for H-H and -0.8896 for each of the O-H, or a total of -1.5912 a.u. The 2-center contribution to the total energy, therefore, compensates for the losses in energy of the molecular orbital valency state (M.O.V.S.) and brings about a binding of 1.5912 a.u. -1.3884 a.u. = 0.2028 a.u. The 3-center energy increases this binding by 0.434 a.u., yielding a total Hartree-Fock binding energy of 0.2462 a.u. (This bond formation process is illustrated in Fig. 4).

Since water has 3 nuclei, clearly we can have only one-, two-, and three centers contributions to the total energy. Let us consider a different set of examples, namely the molecules CH_4, CH_3F, CH_2F_2, CHF_3 and CF_4[41].

We shall now analyze the gross behavior of the M.O.V.S. for the C, H and F atoms in the compounds CH_4, CH_3F, CH_2F_2, CHF_3 and CF_4. Information on the basis set for the wavefunctions of these molecules can be obtained from a recent paper by Siegbahn and Gelius et al.[42]. It is noted that these wavefunctions are somewhat crude approximations of

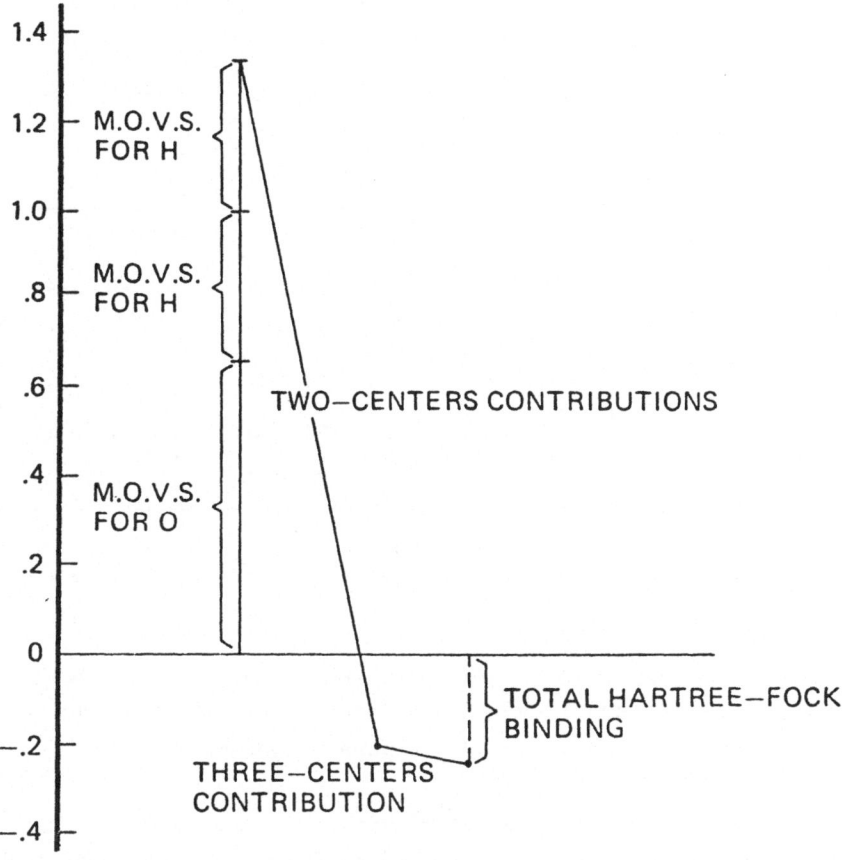

BOND FORMATION DIAGRAM FOR H_2O

Figure 4
Bond formation diagram for the water molecule in the Hartree-Fock approximation. The zero corresponds to the sum of the Hartree-Fock energies of
the component atoms, at disassociation in the corresponding atomic ground
states. The Molecular Orbital Valency States energies (M.O.V.S.) are
given for the oxygen atoms and for the two hydrogen atoms (the scale is in
a.u.). The two- and the three-centers contributions to the total energy
bring about the indicated Hartree-Fock binding.

Hartree-Fock functions; on the other hand, they are sufficiently accurate to indicate trends correctly and, therefore, adopted in this
work; the wavefunctions were re-computed.
In Table 10, we report the M.O.V.S. for the carbon, hydrogen and fluorine atoms.

TABLE 10

M.O. Valency State for the Carbon, Fluorine and Hydrogen Atoms

as from Bond Energy Analysis (in a.u.)*

Molecule	C	F	H
CH_4	−36.6266	---	−0.19942
CH_3F	−36.4193	−99.2982	−0.20531
CH_2F_2	−36.1097	−99.2627	−0.2124
CHF_3	−35.7293	−99.2411	−0.2048
CF_4	−35.3239	−99.2296	---

*Valency State Standard for the Tetrahedral Carbon −37.4585 a.u.

In Table 11, the one-, (E_A), two-, (E(AB)), three-, (E(ABC)), and four-center energies (E(ABCD)), are summed and reported relative to the sum of the Hartree-Fock energy of the separated atoms.

TABLE   11

Decomposition of total energy into one-, two-, three-, four-centers components.

Binding (a.u.)	CH_4	CH_3F	CH_2F_2	CHF_3	CF_4
1 center	+2.2645	+2.0240	+2.3264	+2.7278	+3.0416
1 & 2 center	−0.6405	−1.3331	−1.4850	−1.6026	−1.8127
1, 2 & 3 center	−0.3308	−0.4860	−0.3006	−0.1430	−0.0882
1, 2, 3 & 4 center	−0.5002	−0.6963	−0.5239	−0.3967	−0.3657

These energies (in a.u.) are given relative to the sum of the Hartree-Fock energies of the component atoms in the ground state.

Thus, for example, the one-center sum for CH_3F is reported as E(C) + + 3E(H) + E(F) − E(C, ^3P) − 3E(H, ^2S) − E(F, ^2P). It is noted that E(C, ^3P) = −37.6886 a.u., E(H, ^2S) = −0.5 a.u., E(F, ^2P) = −99.4093 a.u. (see reference 27). Table 12 reports the gross atomic charges for the compounds of this section.

TABLE 12

Gross Charges and Hybridization

	CH_4	CH_3F	CH_2F_2	CHF_3	CF_4
1s (H)	0.826	0.780	0.832	0.823	----
1s (C)	2.000	2.000	2.000	2.000	2.000
2s (C)	1.452	1.346	1.209	1.057	0.910
2p (C)	3.261	2.782	2.435	2.222	2.083
1s (F)	----	2.000	2.000	2.000	2.000
2s (F)	----	1.933	1.942	1.938	1.938
2p (F)	----	5.355	5.404	5.361	5.314

From Table 10, we learn that the M.O.V.S. is strongly dependent on the environment. It ought to be so, and in this respect, the M.O.V.S. seems to be a more useful concept than the V.S.S. The V.S.S. ignores the specific interaction with the neighboring atoms and ignores any charge transfer. By construction, the M.O.V.S. takes into account both effects. In addition, the M.O.V.S. is obtained directly from the Hartree-Fock energy of the molecule in consideration; therefore, it makes an immediate analysis of the Hartree-Fock computed data. A second feature of importance is that the one-center energy by itself gives no binding, and it takes the many-centered energies to get molecular binding. Therefore, we learn (once more) that neglect of three- and four-center integrals in general leads to an incorrect value for the binding energy. (The various techniques whereby the overlap is totally or partially neglected in order to avoid the computation of many-center integrals essentially parameterizes the two-center energy contribution, attempting to obtain not the energy which is associated with the model wavefunction used, but with the experimental binding).
The one-center energy (the one associated with the M.O.V.S.) brings

the molecule in a highly unstable situation, with respect to the energy of the separated atoms; the two-center contribution overcompensates the above effect, and the three- and four-center contribution adjusts the situation to what is known as the Hartree-Fock binding energy. By comparing the data of Table 10 with those of Table 11 (reported graphically in Fig. 5), we learn that the M.O.V.S. energy is nearly a regular function of charge-transfer, and the more an atom acts as an acceptor, the less unstable the M.O.V.S. becomes (relative to the energy of the separated atoms).

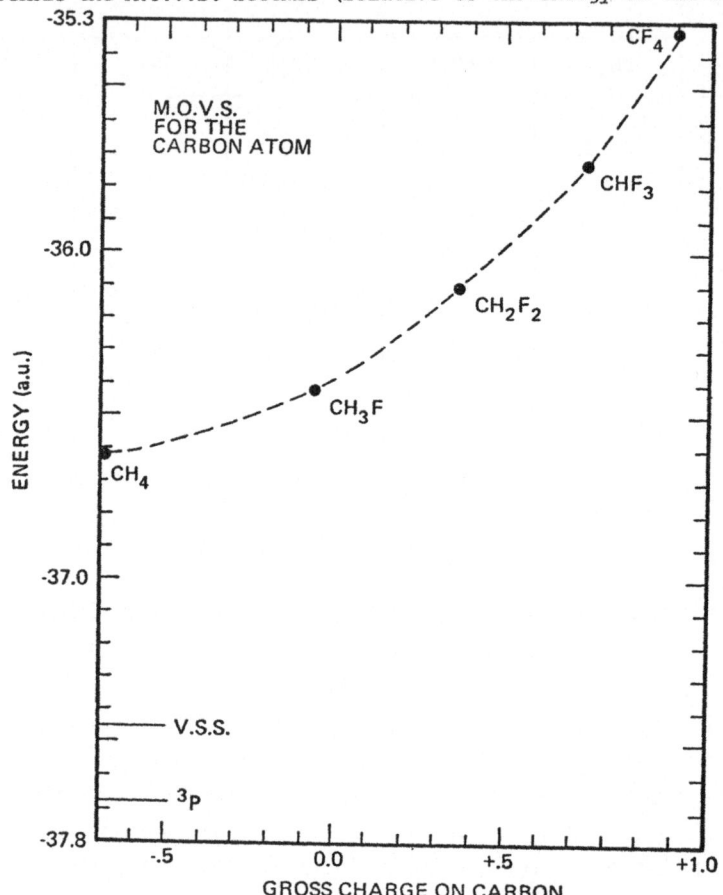

Figure 5 Variation of M.O.V.S. Energy with Environment. The plot is for the energy of the Molecular Orbital Valency State in CH_4, CH_3F, CH_2F_2, CHF_3, CF_4 molecules at their equilibrium configuration. The energy for the carbon atom in the ground state (^3P) and the energy of the Valency State Standard for tetrahedral carbon are given as references. In the abscissa the computed gross charges on the carbon are given. It is noted that the extreme variation of the M.O.V.S. energy is due in part to the restrictions used in the basis set, which manifest themselves in an exaggerated ionic character. Nevertheless, the main features, namely that the M.O.V.S. is much higher than the V.S.S. and is a function of the environment are inherent in the definition of M.O.V.S.

1.11) <u>Bond Energy Analysis for Molecular Complexes</u> - Our assumption to partition the total energy into bond energies requires only the wave-function for a system of m nuclei and n electrons; therefore it clear-ly must hold not only for a separated molecule but also for molecular complexes (or for a system of more than one molecule). Let us consider a water molecule, W, forming a complex W-α where α is a short notation to represent either a cation, like Li^+, Na^+, K^+, or an anion like F^- and Cl^- or a second molecule of water. Useful information for such sys-tems can be obtained considering W and α as a two centers systems, and this is readily done by summing up the one-, two, three-center contributions of water into one single term. In this way the total energy of the above systems can be described by relative few terms; since the total binding energy of the complex (i.e. the total energy of the system minus the energy of the component atoms in the ground state) is a more interesting quantity than the total energy, we shall analyze the quantities:$\Delta E(W, \alpha)$ i.e., the total binding energy of the complex (W, α); $\mathcal{E}_i(W, \alpha)$ i.e., the sum of the bond energies containing matrix elements with at least one basis set or a nuclear charge on the molecule W and on α; $\Delta \mathcal{E}_{1,0}(W)$, i.e. the variation in the energy of the water molecule W due to the field of α, relative to the energy of W unperturbed by α, and $\Delta \mathcal{E}_{1,0}(\alpha)$, i.e. the variation in the energy of α due to the field of W, relative to the energy of α unperturbed by W. (The reason for such subscript can be found in ref. 39). Notice that by definition

$$\Delta E(W, \alpha) = \mathcal{E}_1(W, \alpha) + \Delta \mathcal{E}_{1,0}(W) + \Delta \mathcal{E}_{1,0}(\alpha)$$

In Fig. 6, 7 and 8 we consider $E(W, \alpha)$ for the previously mentioned W-α complexes for a number of distances between the water molecule and the system α, and for the orientation indicated in the figures 6, 7 and 8. In Fig. 6 these quantities are reported for the most stable configuration of $Li^+(H_2O)$, $Na^+(H_2O)$ and $K^+(H_2O)$ as a function of the cation-oxygen distance (R(O-Ion). The total energy curves in Fig. 6a show the decrease in binding energy and increase in cation-oxygen equilibrium separation on going from Li^+ to Na^+ and then to K^+. It can be seen that for R(O-Ion) values greater than about 8 a.u., the energy curves coincide for all three cations (the energies are measured relative to the energy for infinite separation of the cation and water). For large R(O-Ion) values the cation-water interaction can be described by replacing the cation with a point charge of +1. Comparison of the total energies in Fig. 6a with the cation-water in-

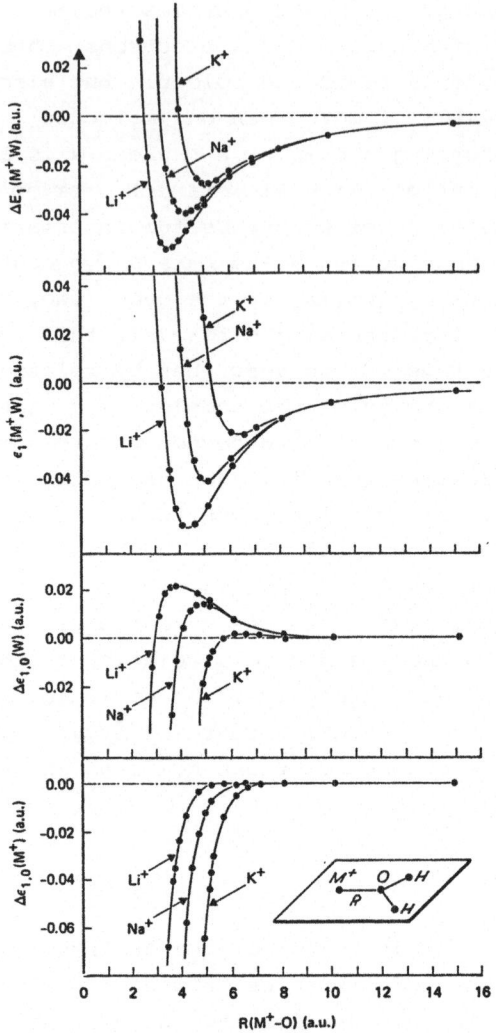

Figure 6 Total energy $\Delta E_1 (M^+, W)$, cation-water interaction energy $\epsilon_1 (M^+-W)$, water energy $\Delta \epsilon_{1,0} (W)$, and cation energy $\Delta \epsilon_{1,0} (M^+)$ for Type 1 configurations (see Fig. 1) as a function of cation-oxygen internuclear separation $R(M^+-O)$. Energies and distances are in a.u. By definition, the sum of $\Delta \epsilon_{1,0} (W)$, $\Delta \epsilon_{1,0} (M^+)$ and $\Delta \epsilon_1 (M^+-W)$ is equal to $\Delta E_1 (M^+, W)$ (see paper I).

teraction energies in Fig. 6b shows that the minima in the $_1$(Ion-W) curves are shifted uniformly to larger R(O-Ion) values in comparison with the minima in the corresponding ΔE_1(Ion, W) curves. On the other hand, the behavior of the energy well depths is quite different. The

ε_1(Ion-W) well depths are larger for $Li^+(H_2O)$ and smaller for $K^+(H_2O)$ than the corresponding well depths for ΔE_1(Ion,W).

In Fig. 6c water energies $\Delta\varepsilon_{1,0}$(W) are compared for the three complexes. It can be seen that the maxima in the curves decrease rapidly on going from Li^+ to K^+. The curves show that the water energy change due to the presence of a cationic field is influenced not only by the nature of the cation but also by its size.

Cation energy changes due to the presence of the water molecule field are shown in Fig. 6d. $\Delta\varepsilon_{1,0}$(Ion) values are approximately the same near the equilibrium cation-oxygen separation for all ions. This indicates that for the equilibrium configuration of the complexes the increase in cationic polarizability on going from Li^+ to K^+ is offset by the increase in cation size. This result is very useful in attempts to use the data obtained for the Li^+, Na^+ and K^+cations to predict the behavior of larger ones like Rb^+ and Cs^+. For these cations the binding energies should be quite similar to that for K^+. The $\Delta\varepsilon_{1,0}$(W) and ε_1(Ion-W) contributions will be fairly small (see Figs. 6b and 6c) and consequently the binding will be mainly determined by the $\Delta\varepsilon_{1,0}$ (Ion) value.

It is of interest to note that for a given complex use of the CNDO approximation leads[43,44] to a total energy whose behavior is intermediate between that for the Hartree-Fock total energies (Fig. 6a) and the cation-water interaction energies (Fig. 6b). This leads to equilibrium R(O-Ion) values that are too large in comparison with experiment. It appears that by neglecting 3- and 4-center integrals in the CNDO approximation the 2-center terms are overemphasized. The latter terms are more important for the cation-water interaction energy than for the total energy.

In Fig. 7, the total Hartree-Fock binding energy (relative to the sum of the Hartree-Fock energies for the separated dissociation products) and the Bond Energy Analysis partitioning are presented for those C_{2v} symmetry and HB (hydrogen bonded) configurations (see Fig. 7 for definitions) for which the geometry of the water molecule was kept constant for the cases of the H_2O-F^- and H_2O-Cl^- complexes[45].

We know that the binding in the C_{2v} and in the HB configurations are nearly the same for the Cl^--H_2O complex, but are very different for the two configurations in the case of the F^--H_2O complex[45]; indeed, this can be seen to be the case from the lowest three diagrams in Fig. 7.

In the following we shall comment on the two different curves for $\Delta\varepsilon_{1,0}$(F^-), when one compares the C_{2v} and the HB configurations. Let

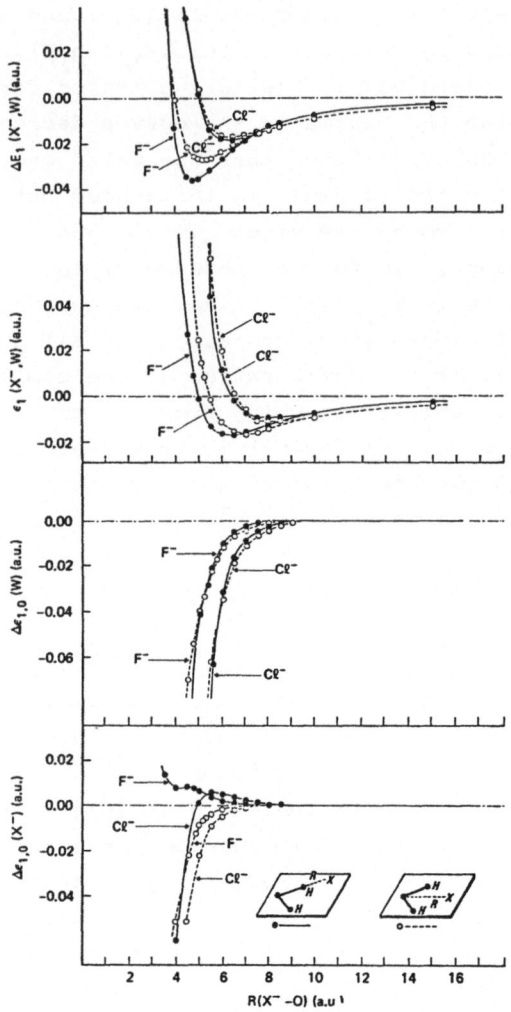

Figure 7 Total Energy and Bond Energy Analysis for the F^- and Cl^- Complexes in the HB and C_{2v} Configurations. The value of the ordinates gives the Total Energy, $\Delta E_1(X^-,W)$ (with zero taken as the energy of the dissociation products, namely the anion and the water molecule), the Interaction Energy of the anion with the water molecule, $\Delta \epsilon_1(X^-,W)$, the energy on the water molecule, the field of the anion, $\Delta \epsilon_{1,0}(W)$ and the energy of the anion in the field of the molecule of water. The curves with empty dots refer to the HB (Hydrogen Bonded) configuration; those with filled dots refer to the C_{2v} configuration.

us start by noting that the net atomic populations obtained using Mulliken's population analysis, one can conclude that there is an appreciable charge transfer from the fluorine to the molecule of water in the C_{HB} configuration[45] but that there is only a small amount of

charge transfer from the fluorine to the molecule of water in the HB
configuration. In addition, in the C_{HB} configuration there is internal
charge transfer from the hydrogen to the oxygen. This internal charge
transfer becomes more pronounced when the F^- approaches the water molecule, and thus the F^- experience a stronger and stronger positive
field. The positive field polarizes the fluorine's electronic density,
leading to a polarization, that increases the anion stability.
In the HB case, there is a small amount of charge transfer from the
anion to the water molecule for decreasing values of the F-O distance:
this effect tends to reduce the stability of the fluorine anion
(as known the F^- ion is more stable than the F atom). The two hydrogens taken together are less positive, with decreased F-O distance
than in the C_{2v} case: for example at $R(F-O) = 5.15$ a.u. the two hydrogens in the C_{2v} configuration have 0.64 of an electron but in the
HB configuration one has 0.60 of an electron and the second 0.74 of
an electron. This tread tends to decrease the F^- stability.
For the case of the H_2O-Cl^- complex, we can easily interpret the
data on Fig. 7, by remembering that the Cl^- ion has a larger radius
than the F^- ion and, threfore, the differences between the C_{2v} and
the HB configurations are relatively small.
In Fig. 8 we show the interaction of two water molecules W, and W_2
for the conformation given on the top of the Figure[46]. Near the
equilibrium position, at an oxygen-oxygen distance of about 5.7 a.u.,
the binding is due totally to polarization if we identify polariza-
tion binding with the quantities $\Delta \varepsilon_{1,0}(W_1)$ and $\Delta \varepsilon_{1,0}(W_2)$. From Fig. 8
we see that there are two distinct polarization patterns as was found
in the case of $F^-(H_2O)$. One water molecule acts in a way analagous
to the F^- anion and the second acts like the H_2O molecule in $F^-(H_2O)$.
Let us designate the water molecule in the XY-plane (Fig. 8) as W_1
and the water molecule in the XZ-plane as W_2; then W_1 is affected
mainly by the negatively charged part of W_2, since the O_2 oxygen is
nearer to W_1 than the W_2 hydrogens. Thus W_1 is analogous to the H_2O
molecule in the $F^-(H_2O)$ complex (in the hydrogen bonded structure designated as HB). W_2 is affected mainly by the positively charged hydrogen of W_1 that forms the hydrogen bond. Thus the energy curve
$\Delta \varepsilon_{1,0}(W_2)$ is expected to be similar to the ones for H_2O in the $M^+(H_2O)$
complexes or for the M^- anion in the $M^-(H_2O)$ complexes.
The water-water interaction term $\Delta \varepsilon_1(W_1,W_2)$ has its minimum substan-
tially shifted from the minimum in the total binding curve $\Delta E_1(W_1,W_2)$.
This was also the case for the positive and negative ions interacting
with water. For these cases one can conclude that the charge rearrange-

44

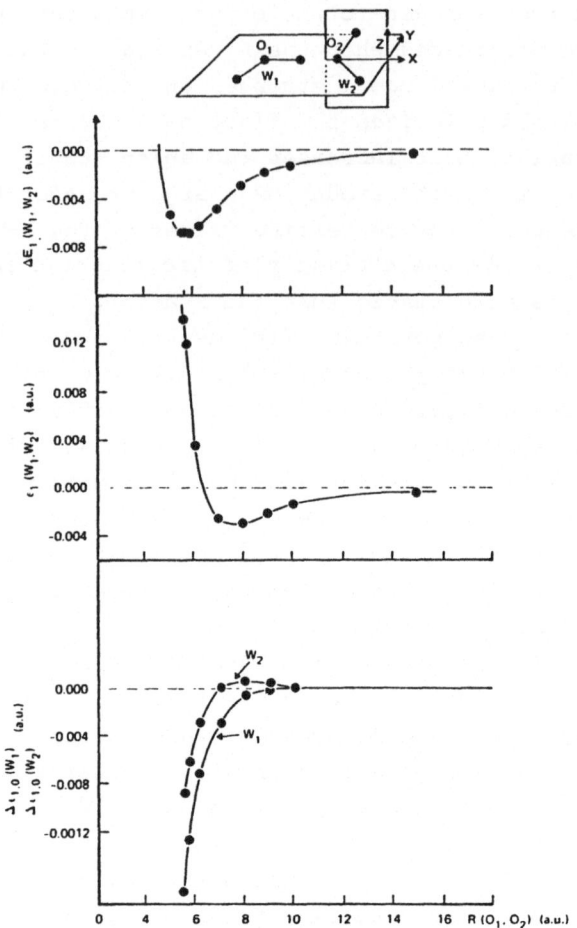

Figure 8 Energy partitioning for the water dimer in the most stable configuration as a function of the oxygen-oxygen separation.

ment (called polarization or internal charge transfer) within the interacting molecules in a hydrogen bonding situation not only accounts for a significant portion of the binding energy, but also is the dominant factor in the determination of the structure of the complex. Because of polarization ($\Delta\varepsilon_{1,0}(W_1)$ and $\Delta\varepsilon_{1,0}(W_2)$ by definition), the water dimer is much more compact that it would be otherwise. Let us pose the question on the accuracy of our computations for the binding energy of the water-ion complexes.

The answer is given below[47]:

Complex of the water with the ions	Li^+	Na^+	K^+	F^-	Cl^-
Hartree-Fock binding energy (kcal/mole)	-35.20	-23.95	-16.64	-27.70	-11.85
Correlation binding (Kcal/mole) [a]	- 0.26	- 0.26	- 0.33	- 0.75	- 0.57
Zero point energy correction (Kcal/mole)	2.24	1.74	1.51	3.40	1.80
Computed H (kcal/mole)	34.13	23.32	16.20	22.18	11.37
Uncertainty in computed value for H[b]	$\pm$ 0.94	$\pm$ 1.00	$\pm$ 0.80	$\pm$ 0.91	$\pm$ 1.70
Experimental H (kcal/mole) [c]	34.0	24.0	17.9	23.3	13.1
Computed Ion-Oxygen distance (in a.u.) [d]	3.58	4.25	5.08	4.75	6.25
Computed Ion-Oxygen-hydrogen angle [d]	180	180	180	47.9	37.8

---

a) estimated with statistical techniques.
b) The uncertainty has been obtained by summing 20% of the zero point energy correction, to 20% of the Wigner-type estimate for the correlation energy, to 50% of the energy difference between the Hartree-Fock binding computed with the large and the small basis set.
c) The experimental error was not reported.
d) The geometry summarized in the last three rows corresponds to the most stable configuration. The angle I is the angle between the C_{2v} axis of the water molecule and the oxygen-ion axis. The rotation direction is defined by setting I=0 for the C_{2v} structure of anions-water complex and by setting I=180 for the C_{2v} structure of the cation-water complex.

Thus we can conclude that our binding energy is remarkably accurate and that the energy partitioning presented in Figs. 6, 7 and 8 must be a realistic representation of the forces responsable for the formation of the W-ion complexes.

1.12) Bond Energy Analyses to Derive Chemical Structural Formulas - In the two previous sections, we have decomposed the total energy of a system using the Bond Energy Analyses in two different (and nearly extreme fashions) either considering each nucleus for each molecule's orbital (analysis of the orbital energies for the water molecule) or grouping all atomic energy terms of a complex (like water-ions or water-water) into three contributions to the total energy, namely: the energy of one molecule (in the field of the second one), the energy of the second molecule (in the field of the first one) and the direct interaction energy between the two molecules.

In this section we shall derive an energy partitioning that is equivalent to the chemical structural formulas and, in addition, by pro-
viding a quantitative value of the energy in each bond, should allow
for a step in the direction of understanding chemical reactivity[48].
The Hamiltonian given in eq. (2), when applied on a wavefunction describing four or more nuclei and the corresponding electrons, yields
the following one-, two-, three- and four centers contributions (see
eq. (37)) to the total energy:

- kinetic part: yields one-, two-center terms;
- nuclear electron attraction part: yields one-, two- and three-center
  terms;
- electron-electron part: yields one-, two-, three- and four-center
  terms;
- nuclear-nuclear repulsion: yields two-center terms.

To map these terms into two-center terms - the necessary condition to
obtain the analogue of the classical chemical formula - we shall make
use of a number of transformations, below described. To simplify the
notation, we shall write a basis set χ_A centered on the atom A simply
as A; with this notation we prescribe the following transformation
rules:

- kinetic matrix elements: the $\langle A|\nabla|A \rangle$ matrix elements are assigned
  to an energy $E(A)$ and the $\langle A|\nabla|B \rangle$ matrix elements are assigned to
  an energy $E(AB)$;
- nuclear-electron matrix elements: the $\langle A/Z_A/R/A \rangle$ elements and the
  $\langle A/Z_B/R/A \rangle$ elements are assigned to the energy $E(A)$; the $\langle A/Z_A/R/B \rangle$
  elements, the $\langle A/Z_C/R/B \rangle$ elements are assigned to the energy $E(AB)$;
- electron-electron matrix elements: the terms $\langle A A/1/r/A A \rangle$ are assigned to the energy $E(A)$; the terms $\langle AA/1/r/A B \rangle$, $\langle AA/1/r/BB \rangle$,
  $\langle AB/1/r/AB \rangle$ are assigned to the energy $E(AB)$; the terms $\langle AB/1/r/CD \rangle$
  are assigned to $E(AC)$, $E(AD)$, $E(BC)$ and $E(BD)$ with the transforma-
  tion relation.

$$\langle AB|1/r|CD \rangle = W(\langle AA|1/r|CC \rangle + \langle AA|1/r|DD \rangle + \langle BB|1/r|CC \rangle + \langle BB|1/r|DD \rangle) =$$

= term (AC) + term (AD) + term (BC) + term (BD), where the weight, W,
  is given by the relation

$$W = \frac{\langle AA|1/r|CC \rangle + \langle AA|1/r|DD \rangle + \langle BB|1/r|CC \rangle + \langle BB|1/r|DD \rangle}{\langle AB|1/r|CD \rangle}$$

- nuclear-nuclear repulsion: the $(Z_A Z_B/R)$ energy is obviously assigned to the E(AB) energy term.

Therefore, with the above transformation rules we can write that the total Energy E corresponding to a wavefunction describing m nuclei and n electrons is decomposed into pairwise energies, or

$$E = \sum_{A} \sum_{B} E(AB) = \sum_{A} E(A) + \sum_{A} \sum_{B \neq A} E(AB) \qquad (41)$$

where the index A (or B) runs over all the atoms of the molecule. Let us now consider a set of four atoms designated as 1, 2, 3 and 4 and let us assume that two of them (1 and 2) form a molecule and the second set (3 and 4) form a second molecule. We consider three conformations for the four atoms: i) where all four atoms are infinitely for one from the other, ii) where atoms 1 and 2 are at a distance where there is a finite non-zero interaction (for example the equilibrium distance) and when the atoms 3 and 4 are at a distance where there is non-zero interaction (for example the equilibrium distance) but such that the two molecules are infinitely separated and iii) when we are as in the previous case but the two molecules interact one with the second. Using eq. (41) for the three cases we obtain three matrices designated as matrix $\underline{A}$, matrix $\underline{B}'$, and matrix $\underline{C}'$, described below:

Matrix A

	1	2	3	4
1	a(11)	0	0	0
2	0	a(22)	0	0
3	0	0	a(33)	0
4	0	0	0	a(44)

Matrix B'

	1	2	3	4
1	b'(11)	b'(12)	0	0
2	b'(21)	b'(22)	0	0
3	0	0	b'(33)	b'(34)
4	0	0	b'(43)	b'(44)

Matrix C'

	1	2	3	4
1	c'(11)	c'(21)	c'(31)	c'(41)
2	c'(21)	c'(22)	c'(32)	c'(42)
3	c'(31)	c'(23)	c'(33)	c'(43)
4	c'(41)	c'(24)	c'(34)	c'(44)

We note that the matrix $\underline{A}$ is defined as that matrix having all elements but the diagonal equal to zero; the B' matrix is the one having a symmetric number of terms (a(ij) and a(ji)) equal to zero; the matrix $\underline{C}'$ is the one having no pair of symmetric terms equal to zero; in addition $\underline{A}$, $\underline{B}'$ and $\underline{C}'$ are square and symmetric matrices. By subtracting each element a(ij) of $\underline{A}$, from the corresponding element b(ij) of $\underline{B}'$ one obtains the total binding energy of case ii.

By subtracting each element a(ij) of $\underline{A}$ from the corresponding element C'(ij) of $\underline{C}'$ one obtains the total binding energy for case iii).

By subtracting each element a(ij) of $\underline{A}$ one and each element b'(ij) of $\underline{B}'$ from the corresponding element c(ij) of $\underline{C}$ one obtains the interaction energy of the two molecules relative to the two molecules at infinite separation one from another. Clearly, if the four atoms (or n atoms) form not two distinct molecules (as in our example) but a single molecule, then only the matrix $\underline{A}$ and the matrix $\underline{C}'$ are of interest.

The matrices $\underline{A}$, $\underline{B}'$ and $\underline{C}'$ above defined have an immediate physical interpretation, as described; however, such matrices suffer from the fact that the diagonal terms are non-zero and therefore can not be used directly to provide a chemical structural formula.

Let us now define a matrix $\underline{B}$ of elements b(ij), a matrix $\underline{C}$ of elements c(ij) and a matrix IP, of elements designated either as P(ij) or B(ij); the constructions rules are as follows:

$$b(ij) = (1-\delta(ij))\left[b'(ij)-a'(ij)+ \sum_K (b'(KK)/\sum_1 b'(11))b'(kk)\right] \qquad (42)$$

$$c(ij) = (1-\delta(ij))\left[c'(ij)-a'(ij)+ \sum_K (c'(KK)/\sum_1 /c'(11))c'(11)\right] \qquad (43)$$

$$P(ij) = c(ij) - b(ij) \qquad\qquad I(ij) = c(ij) - b(ij) \qquad (44)$$

where the indices i and j are valid for a given range of values for p(ij), for a different one for I(ij).

For our example, we have

Matrix $\underline{B}$

	1	2	3	4
1	0	b(12)	0	0
2	b(21)	0	0	0
3	0	0	0	b(34)
4	0	0	b(43)	0

Matrix $\underline{C}$

	1	2	3	4
1	0	c(12)	c(13)	c(14)
2	c(21)	0	c(23)	c(24)
3	c(31)	c(32)	0	c(34)
4	c(41)	c(42)	c(43)	0

Matrix IP

	1	2	3	4
1	0	p(12)	I(13)	I(14)
2	p(21)	0	I(23)	I(24)
3	I(31)	I(32)	0	p(34)
4	I(41)	I(42)	p(43)	0

The matrix $\underline{B}$, represents the energy of the bonds of the first molecule
(namely 2b(12)) and the energy of the bond of the second molecule
(namely 2b(34)); the matrix $\underline{C}$ represents all the bonds in the sense
of pairwise interactions for the two molecules when interacting. The
matrix IP represents the interaction of the two molecules expressed
as the variation (by polarization) of the energy of the bonds of one
molecule due to the field of second one (p(12) and p(21)), the variation of the energy of the bonds of the second molecule due to the
field of the first one (p(34) and p(43)) and the non bonded interactions denoted as I(ij). Clearly, if the system is one single molecule,
then the matrix IP has no meaning (unless we consider the case of vi-
brations or internal rotations and then $\underline{C}$ might represent the equilibrium position and IP a different geometrical conformation).
If we indicate with $\Sigma\,\mathcal{E}(i,M)$ and $\Sigma\,\mathcal{E}(i,N)$ the sums of the energies
of the atoms constituting the molecules M and N, respectively, with
E(M) and E(N) the total energies of the molecules M and N, respectively, and with E(M,N) the total energy of the complex M-N, then the
following equalities hold:

$$\sum_{i,j=1}^{2} a(ij) = \mathcal{E}(M) \qquad \text{and} \qquad \sum_{i,j=3}^{4} a(ij) = \mathcal{E}(N) \qquad (45)$$

$$\sum_{i,j=i}^{2} b'(ij) = E(M) \qquad \text{and} \qquad \sum_{i,j=3}^{4} b'(ij) = E(N) \qquad (46)$$

$$\sum_{i,j} c'(ij) = E(M,N) \qquad (47)$$

The binding energy of M expressed as the sum of all bonded and non
bonded interactions is

$$\sum_{i,j=1}^{2} b(ij) = E(M) - \mathcal{E}(M) \qquad (48)$$

The binding energy of N expressed as the sum of all the bonded and non bonded intéractions is

$$\sum_{ij=3}^{4} b(ij) = E(N) - \mathcal{E}(N) \qquad (49)$$

The binding energy of the complex M-N expressed as the sum of all the bonded and non bonded interactions is

$$\sum_{ij} c(ij) = E(M,N) - E(M) - A(N) \qquad (50)$$

The polarization energy of M due to N is $\sum_{ij=1}^{2} p(ij)$ and the polarization energy of N due to M is $\sum_{ij=3}^{4} p(ij)$ and the sum of non bonded interactions between M and N in the complex is $2 \sum_{i=1}^{2} \sum_{j=3}^{4} I(ij)$.

The extension to more than two molecules is trivial. It is stressed that these definitions hold not only for any function in the SCF-LCAO-MO approximation, but also for C.I. wavefunction or any wavefunction expressed as a linear combination of Slater determinants.

From the $\underline{C}$ matrix we can obtain a chemical structural formula: the rule is simple; all positive matrix elements $c(ij)$ represent repulsive energies between atom i and atom j; all negative matrix elements $c(ij)$ represent attractive energies between atom i and atom j. A classical bond is identified as all those $c(ij)$ matrix elements having negative energies and for atoms i and j that are nearest neighbors. Thus we have accomplished the task to obtain structural formulas from a wavefunction constructed by considering a system as a set of m fixed nuclei and n electrons. But we have more; from the non bonded interactions we know those that are attractive and those that are repulsive and for bonded and non bonded interaction we have a numerical quantity defining the bond strength.

Let us now present as a numerical example the case of CH_4+H_2O complex[49]; the oxygen atom is placed on a C-H line, at an oxygen carbon distance of 6.0 a.u. with the two hydrogens atoms symetrically disposed on the C-O axis (that is therefore made to coincide with the C_2 symmetry axis of H_2O) and pointing away from the CH_4 molecule. For this example we use a minimal gaussian basis set (3 s-type functions for each hydrogen, 7 s-type, 3 $2p_x$-type, 3 $2p_y$-type and 3 $2p_z$-type gaussians for the oxygen and for the carbon atoms). This is therefore a system of 8 nuclei (at the above fixed positions) and 20 electrons. The matrices $\underline{A}$, $\underline{B}'$, $\underline{C}'$, $\underline{B}$, $\underline{C}$ and IP are given in Table 13.

To save space in presenting Table 13, the atoms in CH_4 and in H_2O are identified by a number from 1 to 8 and the correspondence between atom and number is explicitly given only once (Table 13a). In addition, we report only the lower part of the matrix, since the matrices $\underline{A}$, $\underline{B}'$, $\underline{C}'$, $\underline{B}$, $\underline{C}$ and $\underline{IP}$ are symmetrical.

With the previously defined basis set the $\underline{A}$ matrix (atomic energies for the atoms at infinite separation) is as given below:

TABLE 13a

Matrix $\underline{A}$ for the $CH_4 + H_2O$ study (in a.u.)

	CH$_4$					H$_2$O		
	C	H	H	H	H	O	H	H
	1	2	3	4	5	6	7	8
1	-37.609							
2	0.0	-0.497						
3	0.0	0.0	-0.497					
4	0.0	0.0	0.0	-0.497				
5	0.0	0.0	0.0	0.0	-0.497			
6	0.0	0.0	0.0	0.0	0.0	-74.621		
7	0.0	0.0	0.0	0.0	0.0	0.0	-0.497	
8	0.0	0.0	0.0	0.0	0.0	0.0	0.0	-0.497

Note: these values correspond to the basis set described; the correct Hartree-Fock energies are: -0.500 a.u. for H('S), -37.688 a.u. for C(^{3}P) and -74.809 a.u. for O(^{3}P).

In Table 13b we report the matrix $\underline{B}'$, namely the matrix describing the two molecules CH_4 and H_2O at infinite separation. From this matrix we learn, once more, that the molecular orbital valency state energies (i.e. the energy differences between the isolated atom in the ground state and the atom "in the molecule") are large and repulsive. We learn, by inspection, that the bonds are between the carbon and the hydrogens in CH_4, between the oxygen and the hydrogens, in H_2O, and all the remaining pairwise interactions are small and repulsive (positive sign).

TABLE 13b

Matrix $\underline{B}'$ for the $CH_4 + H_2O$ study (in a.u.)

	1	2	3	4	5	6	7	8
1	0.997							
2	-0.841	0.291						
3	-0.841	0.068	0.291					
4	-0.841	0.068	0.068	0.291				
5	-0.841	0.068	0.068	0.068	0.291			
6	0.0	0.0	0.0	0.0	0.0	0.433		
7	0.0	0.0	0.0	0.0	0.0	-0.692	0.294	
8	0.0	0.0	0.0	0.0	0.0	-0.692	0.191	0.294

In Table 13c we report the matrix $\underline{C}'$, namely the matrix describing
the two molecules CH_4 and H_2O at finite separation (therefore interacting). By comparing Table 13b with Table 13c we learn that the interaction energies of CH_4 with H_2O are smaller in value than the bonding
energies and some are attractive (the C of CH_4 with the H of H_2O, the
O of H_2O with the H of CH_4) and some repulsive (the H of H_2O with the
H of CH_4). We learn that the interaction is stronger for pairs of atoms
that are nearer, as expected; finally we learn that the field of H_2O
on CH_4 brings about small variations in the pairwise interactions describing CH_4 (and that the field of CH_4 on H_2O brings about small variations in the pairwise interactions describing H_2O). To this qualitative information, Table 13c provides in addition quantitative data.
In Table 13d, we report the matrix $\underline{B}$, namely the matrix describing the
two molecules CH_4 and H_2O at infinite separation, after having performed
the transformation described by eq. (42) (i.e. after partitioning the
diagonal elements ónto the non-diagonal elements). This matrix is the
one to be used to obtain the structural chemical formulas. It provides
with the equivalent information of Table 13b, but this is presented in
a more classical way, since the diagonal elements are "adsorbed" into
the non-diagonal.

TABLE 13c

Matrix $\underline{C}'$ for the CH_4+H_2O study (in a.u.)

	1	2	3	4	5	6	7	8
1	0.983							
2	-0.844	0.309						
3	-0.821	0.063	0.285					
4	-0.826	0.063	0.639	0.285				
5	-0.826	0.063	0.639	0.864	0.285			
6	0.117	-0.068	-0.021	-0.021	-0.021	0.445		
7	-0.042	0.018	0.008	0.009	0.009	-0.700	0.280	
8	-0.042	0.018	0.009	0.009	0.009	-0.700	0.190	0.280

TABLE 13d

Matrix $\underline{B}$ for the CH_4+H_2O study (in a.u.)

	1	2	3	4	5	6	7	8
1	0.0							
2	-0.253	0.0						
3	-0.253	0.104	0.0					
4	-0.253	0.104	0.104	0.0				
5	-0.253	0.104	0.104	0.104	0.104			
6	0.0	0.0	0.0	0.0	0.0	0.0		
7	0.0	0.0	0.0	0.0	0.0	-0.215	0.0	
8	0.0	0.0	0.0	0.0	0.0	-0.215	0.312	0.0

In Table 13e, we report the matrix $\underline{C}$, namely the matrix describing the two molecules CH_4 and H_2O at finite separation, after having performed the transformation described by eq. (43) (that partitions the diagonal elements on the non-diagonal). Again the bonds, in classical sense, can be identified by inspections simply by selecting the pair of atoms with corresponding negative sign and nearest neighbors. In addition, Table 13e provides in a very simple form the attractive and the repulsive pairwise interaction either within a single molecule or between the two molecules.

TABLE 13e

Matrix $\underline{C}$ for the CH_4+H_2O study (in a.u.)

	1	2	3	4	5	6	7	8
1	0.0							
2	-0.277	0.0						
3	-0.275	0.094	0.0					
4	-0.278	0.094	0.096	0.0				
5	-0.278	0.094	0.096	0.103	0.0			
6	0.203	-0.028	-0.009	-0.009	-0.009	0.0		
7	-0.012	0.028	0.013	0.013	0.013	-0.271	0.0	
8	-0.012	0.028	0.013	0.013	0.013	-0.271	0.303	0.0

Finally, in Table 13f, we report the matrix $\underline{IP}$, namely the matrix that describes how the pairwise interaction in CH_4 are "polarized" by the presence of H_2O, and how the pairwise interaction in H_2O are "polarized" by the presence of CH_4; Table 13f provides in addition the pairwise interactions between CH_4 and H_2O (see eq. (44)). The matrix elements of the table are very informative: the presence of H_2O stabilizes by 0.02 a.u. the C-H bonds in CH_4 and increases the hydrogen-hydrogen repulsion in CH_4; the presence of CH_4 stabilizes by 0.06 a.u. the O-H bonds in H_2O and decreases the hydrogen-hydrogen repulsion (the pairwise interactions between atoms in the two molecules have been previously discussed).

TABLE 13f

Matrix $\underline{IP}$ for the $CH_4 + H_2O$ study (in a.u.)

	1	2	3	4	5	6	7	8
1	0.0							
2	-0.024	0.0						
3	-0.022	0.009	0.0					
4	-0.025	0.009	0.008	0.0				
5	-0.025	0.009	0.008	0.001	0.00			
6	0.203	-0.028	-0.009	-0.009	-0.009	0.0		
7	-0.012	0.028	0.013	0.013	0.013	-0.057	0.0	
8	-0.012	0.028	0.013	0.013	0.013	-0.057	-0.010	0.0

We hope that by the detailed analysis of the interaction of water with water, water with ions, and water with methane, we have given a suffi-cient indication that a great deal of information is available from quantum mechanical computations. We hope, in addition, to have clearly established that by starting from a theoretical frame work that consid-ers nuclei and electrons, we can arrive in a very natural way to the traditional concepts and to the traditional representations of mole-cules.

We conclude this section with few examples of additional uses and ap-plications for the bond energy analyses.

a) Reaction mechanisms - Reaction paths are often studied with quantum mechanical methods. The use of the "binding energies" in addition to the use of "orbital energies" and the use of the $\underline{B}$ and $\underline{C}$ matrices should provide a detailed and localized picture of the energy changes that follow a reaction[49].

b) Reactivity indices - The influence of all bonds in a molecule or in a reactive complex on a given bond of the molecule (or of the reactive complex) can easily be obtained from the $\underline{C}$ matrix, for example, by ad-ding to the bond energy for the i-j pair of atoms the energy of all the remaining pairs and by introducing an appropriate bond to bond distance relationship. In this way to each pair we can associate an index

(obtained as above described) that from preliminary studies corre-
lates nicely with the probability of chemical substitutions at that
bond[49].

c) Computer oriented studies of organic syntheses - Ugi's brilliant
attempts to obtain an appropriate algebraic frame work in order to
make a step towards computer oriented syntheses for organic chemistry,
has been in practice deterred by the inconvenience of generating too
many reaction paths with no way to critically eliminate the less probable one, if not by recurring to semi-empirical rules, that do not
derive from the algebraic framework[50]. The use of the matrices B and
C seems to us to be the logical way to represent two molecules during
interaction, and the reactive indices, above mentioned, seem to offer
an easy way to select among the many reaction paths, those most probable on an energetic base (studies are in progress along these
lines[49]).

d) Vibrational analyses - The matrices B and C are those that describe
a central force field, therefore are a most adapted quantum chemical representation to obtain force constants and frequencies from ab-
initio computation. As known at present force-constants and vibrational frequencies are obtained by an analysis of the total energy computed for many configurations, rather than by the use of the matrices
B and C[51].

e) Chemisorption and catalysis - When a molecule approaches a surface,
the total energy variation of the complex surface + molecule can be
followed by ab-initio computation. However, the total energy variation
by itself cannot provide any specific information of the interaction
between the atoms of the molecule and the atoms of the surface. However, with the bond energy the situation is different. For example, a
set of graphite rings interacting with a molecule of CO provides information on the specific energy of interactions of the carbon atoms at
the border (surface state) and of those inside, and (if more of one
layer of graphite is considered) of those in the bulk. Therefore, by
considering a sufficiently large cluster of atoms of carbon arranged
according the geometry of graphite, the mechanisms of adsorption can
be elucidated[52].

This list is only indicative, but we hope it is sufficient to interest
the reader in the use of the bond energy, and in stimulating to a search
of alternative ways to decompose the total energy into bond energies.

1.13) <u>References</u>

1) L. C. Pauling, Nature of Chemical Bond, Cornell University Press, Ithaca, New York (1960).

2) E. Clementi, J. Chem. Phys. <u>46</u>, 3851 (1967).

3) R. S. Mulliken, fo example, "Electronic Structure of Molecules", J.C.P. <u>3</u>, 375 (1935).

4) C. A. Coulson, Valency, Oxford University Press (1959).

5) E. V. Condon and G. H. Shortley, The Theory of Atomic Spectra, University Press, Cambridge (1957). See, in addition, J. C. Slater, Theory of Atomic Structure (Mc Graw-Hill Book Co., Inc., New York) Vol. I e Vol. II.

6) C. C. J. Roothaan, Rev. Mod. Phys. <u>23</u>, 69 (1951). See, in addition, D. R. Hartree, The Calculation of Atomic Structures (John Wiley & Sons, Inc., New York (1957).

7) A. C. Wahl, J. Chem. Phys. <u>41</u>, 2600 (1964).

8) V. Fock, Invest. Akad. Nauk. USSR, Ser. Fiz. <u>18</u>, 161 (1954).

9) A. C. Hurley, J. E. Lennard-Jones and J. A. Pople, Proc. Roy. Soc. (London) <u>A220</u>, 446 (1953).

10) E. Wigner, Phys. Rev. 46, 1002 (1934) and E. Wigner and F. Seitz, Phys. Rev. <u>43</u>, 804 (1933).

11) P. O. Löwdin, Advances in Chemical Physics, Vol. 2., Ed. I. Prigogine, Intersciences Publishers, New York (1959).

12) E. Clementi, J. Chem. Phys. <u>38</u>, 2248 (1963).

13) P. Gombas, Pseudopotentiale, Springer-Verlag. New York (1967).

14) E. Clementi and C. Salez., Correlation Energy in Atomic Systems VI. (unpublished results).

15) E. Clementi, J. Chem. Phys. <u>39</u>, 175 (1963); <u>42</u>, 2783 (1965); E. Clementi and A. Veillard, J. Chem. Phys. <u>44</u>, 3050 (1966); <u>49</u>, 1300 (1967).

16) E. A. Hylleraas, Zeits. f. Physik, <u>48</u>, 469 (1928). See, in addition, H. Bethe, Zeits. f. Physik 47, 815 (1929).

17) S. F. Boys, Proc. Roy. Soc. (London), <u>A200</u>, 542 (1950).

18) S. F. Boys, Proc. Roy. Soc. (London), <u>A201</u>, 125 (1950).

19) D. R. Hartree, W. Hartree and B. Swirles, Phil. Trans. Roy. Soc. (London) <u>A238</u>, 223 (1939).

20) A. P. Yutsis, Zh. Eksperim. i. Teoret. Fiz. <u>23</u>, 129 (1952); ibid. <u>24</u>, 425 (1954); A. P. Yutsis, Soviet Phys. -JETP 2, <u>481</u> (1956). See, in addition, T. L. Gilbert, J. Chem. Phys. <u>43</u>, S248 (1956); A. P. Yutsis, Ya. I. Vizbaraire, T. D. 'Strotskire and A. A. Band-

zaitis, Optics and Spectroscopy 12, 83 (1962); A. Veillard and
E. Clementi, Theoret. Chim. Acta, 7, 133 (1967); G. Das and A. C.
WahlJ. Chem. Phys. 44, 87 (1966)E. Clementi and A. Veillard,
J. Chem. Phys. 44, 3050 (1965).

21) O. Sinanoglu, J. Chem. Phys. 36, 706 (1962).

22) B. O. Ross, Chem. Phys. letters 15, 153 (1972).

23) B. O. Ross and P. E. M. Siegbahm "The CIMI Method" to be published
(private communications).

24) G. C. Lie and E. Clementi, J. Chem. Phys. 60, 1275 (1974).

25) G. C. Lie and E. Clementi, J. Chem. Phys. 60, 1288 (1974).

26) E. Clementi, J. Chem. Phys. 35, 33 (1962).

27) First Row Neutral Atoms and Positive Ions - E. Clementi, J. Chem.
Phys. 38, 2248 (1963). Second Row Neutral Atoms and Positive Ions-
E. Clementi, J. Chem. Phys. 39, 175 (1963); third Row Neutral Atoms
and Positive Ions - E. Clementi, J. Chem. Phys. 42, 2783 (1965);
fourth Row Neutral Atoms - C. Roetti and E. Clementi, J. Chem.
Phys. 60, 3342 (1974); 60, 9725 (1974). E. Clementi, "Tables of
Atomic Functions", IBM J. Res. Develop. Suppl. 9, 2 (1965).

28) E. Clementi and C. Roetti, Atomic Data and Nuclear Data Tables,
Volume 14, No. 3 and No. 4, Academic Press, N. Y. (1974).

29) F. Van Duijneveldt, IBM Research Report R.J. 945 (1971).

30) R. S. Mulliken, J. Chem. Phys. 23, 1833, 1841, 2338, 2343 (1955).

31) E. Clementi, J. Chem. Phys. 36, 36 (1962).

32) F. Cavallone, R. Scordamaglia and E. Clementi, An analytical potential obtained from ab-initio computations to represent the interactions of Glycine, Alanine, Valine, Leucine and Isoleucine
with water, J. Am. Chem. Soc. (submitted).   R. Barsotti, F. Cavallone, R. Scordamaglia and E. Clementi, An analytical potential
from ab-initio computations to describe the interaction energy of
Arginine, Asparagine, Lysine, Glutamine, Glutamic acid and Aspartic acid with water, Theoretica Chimica Acta (submitted).   C. Tosi, F. Cavallone, R. Scordamaglia and E. Clementi, An analytical
interaction potential from ab-initio Computations to represent
Proline, Hydroxyproline and Hystidine, Z. Physik. Chemie, N. F.
(submitted).   A. Martellani, F. Cavallone, R. Scordamaglia and
E. Clementi, Ab-initio computations to represent the interaction
of tryptophan, Tyrosine, Threonine and Serine with water, Biochim.
Biophys. Acta (submitted).   R. Pavani, F. Cavallone, R. Scordamaglia and E. Clementi, An analytical potential obtained from ab-
initio computations to represent the sulphur containing amino
acids: Cystine, Cysteine and Methionine interacting with water,

J. Chem. Soc., Faraday Transactions II (submitted).

33) D. Ferro and E. Clementi (to be published).

34) J. Moult, A. Yonath, W. Traub, A. Smilansky, A. Podjarny, D. Rabi-
novich and A. Saya, J. Mol. Biol. (in press).

35) The data on the lysozime dipole moment and Fig. 3 are part of an
unpublished work by G. Ranghino, R. Scordamaglia, G. Giunchi and
E. Clementi.

36) W. Heitler and G. Rumer, Z. Physik. 68, 12 (1931). See, in addition, J. H. Van Vleck, J. Chem. Phys. 1, 219 (1933).

37) R. S. Mulliken, J. Chem. Phys. 2, 782 (1934).

38) H. H. Voge, J. Chem. Phys. 4, 581 (1936).

39) H. Popkie and E. Clementi, J. Chem. Phys. 57, 1077 (1972).

40) E. Clementi, J. Chem. Phys. 47, 4485 (1967).

41) E. Clementi and A. Routh, Int. J. Quantum Chem., Vol. VI, 525
(1972).

42) U. Gelius; P. H. Heden, J. Hedman, B. J. Lindberg, R. Manne, R.
Nordberg, C. Nordling and K. Siegbahn, Molecular Spectroscopy by
Means of E.S.C.A., V.U.I.P. 714, July 1970. Uppsala University,
Department of Physics.

43) H. Liscka, T. Plesser and P. Schuster, Chem. Phys. Letters, 6,
263 (1970).

44) K. G. Breitschwerdt and H. Kistenmaker, Chem. Phys. Letters, 14,
288 (1972).

45) H. Kistenmacher, H. Popkie and E. Clementi, J. Chem. Phys. 58,
5627 (1973).

46) H. Popkie, H. Kistenmacher and E. Clementi, J. Chem. Phys. 59, 3
(1973).

47) H. Popkie, H. Kistenmacher and E. Clementi, J. Chem. Phys. 59,
(1973).

48) E. Clementi, R. Scordamaglia and G.C. Lie, to be published.

49) E. Clementi and R. Scordamaglia, to be published.

50) J. Gastaiger, P.D. Gillespie, D. Marquarding and I. Ugi, Topics
in Current Chemistry, No. 48, Springer-Verlag (Berlin, 1973).

51) E. Clementi, R. Scordamaglia and A. Zerbi, work in progress.

52) M. Giunchi, E. Clementi, work in progress.

Part 2 Structure of Liquid Water as a Test Case

2.1 Introduction - In the previous chapter we have analyzed how a wave-function can be used to describe either a single molecule of water or a molecule of water interacting with ions or with a second molecule of water. In particular, we have stressed the chemical and physical interpretation of the quantities such as electronic density, total energy and orbital energies that constitute the traditional output of a quantum mechanical computation. In this section we shall be confronted with another problem: how to describe a system of many molecules such as a liquid. Clearly, if we fail to describe a liquid as water, there are reasonable doubts on the possibility to describe the structure of an ion or of a molecule when surrounded by the many molecules of water considered as a solvent. Therefore, our aim in this chapter is to produce the structural information (X-rays and neutron diffraction) today available for liquid water, considered as a solution composed of one molecule of water surrounded by many molecules of water. In so doing, we shall find the need of another quantity, previously not even mentioned, namely the temperature and (as its consequence) several aspects of statistical thermo-dynamics. Therefore we shall have a new "many body systems" with reference no longer to electrons and nuclei but molecules and ions, and accordingly we shall pass from the Fermi statistics to the Boltzmann statistics.

2.2 Approximate Hartree-Fock Potential for the Water Dimer - In this section we report a study of the Hartree-Fock potential for the water dimer[1] obtained with the restriction that the two monomers are kept rigid (each with the experimental H_2O geometry; the calculations were carried out using the H_2O basis set given in reference[2]).

We note that the hydrogen bond strength in the water dimer is about 5 kcal/mole (0.008 a.u.). Thus particular care should be used in selecting the basis set. An insufficiently small basis set will yield different accuracy for different water-water separations. At shorter distances the basis set of one molecule will tend to compensate for the deficiencies in the basis set of the other molecule, thereby yielding a lower energy and a larger biding energy[3]. To ensure that our results are of sufficient accuracy, we have carried out a calculation near the equilibrium configuration using a very large basis set that yields a total energy close to the Hartree-Fock limit.

Calculations on the water dimer have been reported[3] for 190 different geometries. As noted previously, the geometry of each water molecule in the dimer is kept rigid and the first water molecule is placed in a fixed position. It lies in the xy-plane with its symmetry axis along the x-axis. The second water molecule is moved around the first one.

In the vicinity of the energy minimum an additional 26 points on the energy surface have been calculated. We have attempted to find an analytical expression for the energy surface by fitting the 216 calculated points. As a starting point we tried several models proposed for the H_2O-H_2O potentials, [4,5,6]. The analytical formula that best combines numerical accuracy with mathematical simplicity can be considered as resulting from an H_2O charge distribution similar to that proposed by Bernal and Fowler[4]. We wish to stress that the point charge model is only a convenient way to simplify the choice of an analytical expression for the potential. We attach no significance to the values of the numerical parameters of the point charge model given in Fig. 1. The parameters obtained are the ones that best fit our numerical Hartree-Fock data.

The analytical potential that we obtained is given by the expression (in a.u.):

$$E = q^2 \left(1/r_{13} + 1/r_{14} + 1/r_{23} + 1/r_{24}\right)$$
$$+ 4q^2/r_{78}$$
$$- 2q^2 \left(1/r_{18} + 1/r_{28} + 1/r_{37} + 1/r_{47}\right)$$

$$+ a_1 \exp(-b_1 r_{56})$$
$$+ a_2 [\exp(-b_2 r_{13}) + \exp(-b_2 r_{14}) + \exp(-b_2 r_{23}) + \exp(-b_2 r_{24})]$$
$$+ a_3 [\exp(-b_3 r_{16}) + \exp(-b_3 r_{26}) + \exp(-b_3 r_{35}) + \exp(-b_3 r_{45})]$$

where: $a_1 = 582.277054$; $a_2 = 0.143789$; $a_3 = 5.470184$; $b_1 = 2.520593$; $b_2 = 1.221756$; $b_3 = 1.936626$; $q^2 = 0.449387$ and the O_5-M_7 (or O_6-M_8) distance is 0.436 a.u. The labelling of the water dimer nuclei and M points is given in Fig. 1.

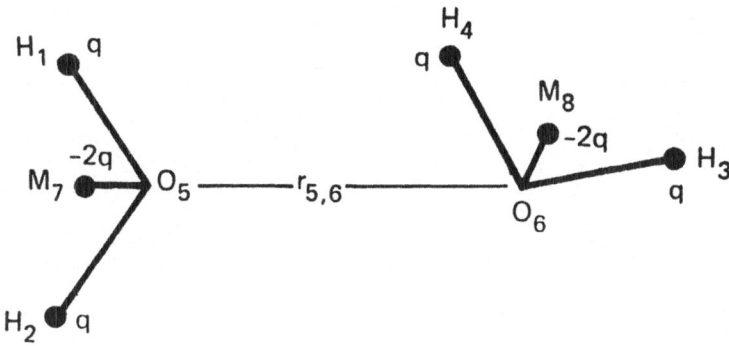

Figure 1   Point charge model used in the derivation of the analytical water-water interaction potential.

The accuracy of this fit is indicated by the following values of the standard deviation. The standard deviation is 0.0002 a.u. for the 80 points with negative E (i.e., that are binding), 0.0004 a.u. for the 165 points with E less than $^+5$ kcal/mole, 0.0013 a.u. for the 208 points with E less than 1 eV, and 0.0018 a.u. for all 216 points.

Thus our fit is very good for attractive regions of the potential, but is somewhat less accurate for the repulsive regions. Of course this is what we want, since repulsive configurations are not important (because of the Boltzmann factor) in the study of systems at room temperature.

The Hartree-Fock potential can be compared with the empirical potentials proposed by Rowlinson[5] and Ben-Naim and Stillinger[6]. We note that these authors were mainly interested in deriving an effective pair potential for the condensed phase rather than a potential for the water dimer.

It is of particular interest to present the energy surfaces where the constraint on the orientation of the symmetry axis of molecule 2 is relaxed. For a given oxygen-oxygen distance, the symmetry axis of molecule 2 is positioned so that the dimer energy is at its minimum value. The resulting surfaces are given in Fig. 2, 3 and 4. All three potential show nearly linear hydrogen bonds for the most stable configurations. The oxygen-oxygen equilibrium separation decreases on going from Fig. 2 to Fig. 4. By counting contours one determines that the Hartree-Fock minimum is not as deep as that for the empirical potentials. In addition, for the empirical potential the hydrogen bonding is more localized near the potential minimum. The linear hydrogen bond orientations are much more favoured than the orientations where the symmetry axes of the two water molecules are parallel.

The optimum angle between the line joining the two oxygen nuclei and the symmetry axis of molecule 2 is shown in the right half of Figs. 2-4. The qualitative behaviour is similar for all three potentials. For the Hartree-Fock potential, in a given region of space there is hardly any reorientation of the symmetry axis of molecule 2, i.e., the value of α is approximately zero. This is less true for the Rowlinson potential and even

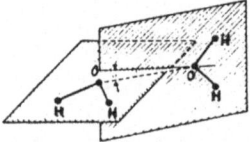

H$_2$O – H$_2$O Hartree - Fock Potential

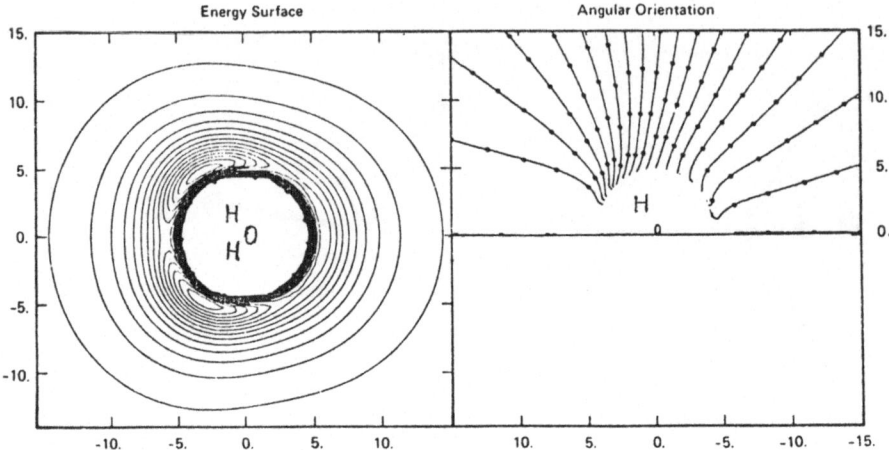

Figure 2 Hartree-Fock potential surface for the illustrated configuration type as a function of the oxygen-oxygen distance and angle α between the symmetry axes of molecule 2 and the line joining the two oxygen nuclei. For the plot on the left, the contour interval is 0.0005 a.u. The oxygen atom is at the origin and the distance scale is in a.u. For the plot on the right, the contour interval is 10°. The horizontal line to the left of the origin is the 0° contour. Only the upper part of the symmetric plot is given.

H$_2$O – H$_2$O Rowlinson Potential

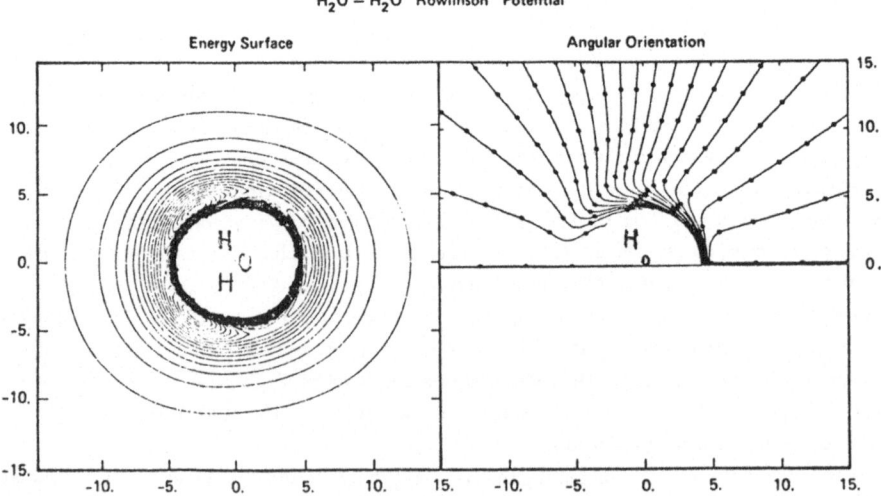

Figure 3 Rowlinson potential surface for the illustrated configuration type

TABLE 1   Water Dimer Binding Energy Computed by Various Authors.

Authors	Uncontracted Basis Set[a]	Contracted Basis Set[a]	H$_2$O Energy (a.u.)	(H$_2$O)$_2$ Energy (a.u.)	(H$_2$O)$_2$ Binding Energy (Kcal/mole)
Morokuma & Pedersen (1968) (α = 0°)	(5,3/3)	(5,3/3)	-75.54939	-151.11887	12.6
Kollman & Allen (1969) (α = 25°)	(10,5/5)	(3,1/1)	-75.97468	-151.95768	5:25
Del Bene & Pople (1970) (α = 58°)	(8,4/4)	(2,1/1)	-75.50013	-151.00998	6.09
Hankins, Moskowitz & Stillinger (1970) (α = 40°)	(10,5,1/4,1)	(5,3,1/2,1)	-76.0458	-152.09069	4.72
Diercksen (1971) (α = 0°)	(11,7,1/6,1)	(5,4,1/3,1)	-76.05326	-152.11167	4.83
Present work (α = 0°)	(11,7,1/6,1)	(4,3,1/2,1)	-76.05525	-152.11747	4.37
(α = 30°)[b]	(11,7,1/6,1)	(4,3,1/2,1)	-76.05525	-152.11784	4.60
Present work (α = 0°)	(11,7,2/6,2)	(4,3,2/2,2)	-76.05990	-152.12561	3.65
(α = 0°)	(13,8,1,1/6,1,1)	(8,5,1,1/4,1,1)	-76.06255	-152.13206	4.37
(α = 30°)[b]	(13,8,1,1/6,1,1)	(8,5,1,1/4,1,1)	-76.06255	-152.13244	4.60
Present work (α = 0°)[b]	(13,8,2,1/6,2,1)	(8,5,2,1/4,2,1)	-76.06598	-152.13781	3.67
(α = 30°)[c]					3.90

(a) The notation (13,8,2,1/6,2,1) indicates that the O atom basis set consists of 13 s-, 8 p-, 2 d- and 1 f-type Gaussian and the H atom basis set consists of 6 s-, 2 p- and 1 d-type Gaussians.

(b) For the definition of α see Table I.

(c) Extrapolated value.

less for the Ben-Naim-Stillinger potentials. This results from the pronounced tendency of the empirical potentials towards nearly tetrahedral orientation of neighboring molecules.

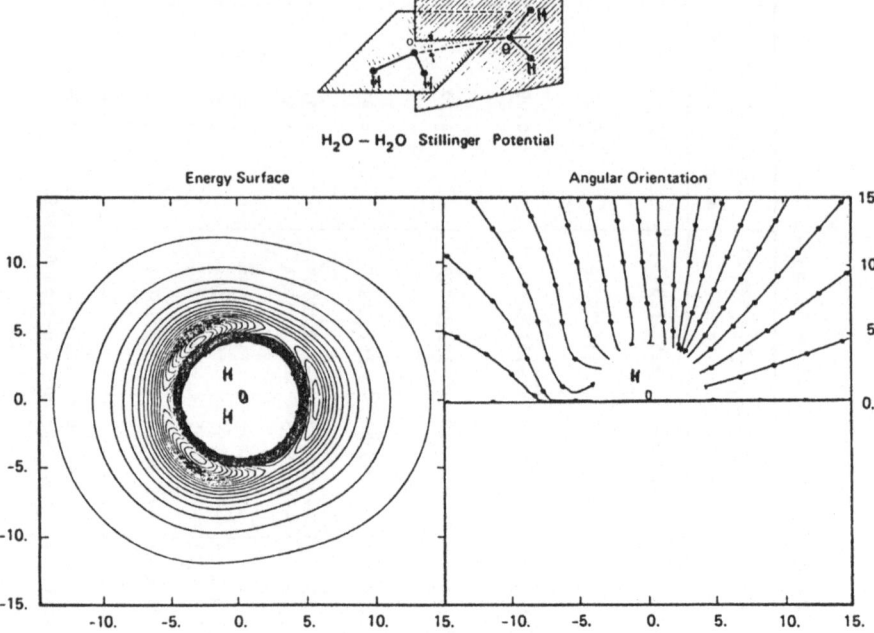

Figure 4 Ben-Naim and Stillinger potential surface for the illustrated configuration type

In Table 1 we compare a number of computations in the SCF-LCAO-MO apporximation to determine the binding energy of the water dimer; those of Diercksen et al. and Hankins et al. correspond to references 7 and 8 respectively.

2.3 Polymers of Water

The more a potential for two molecules is spherical, likely the less meaningful, in a thermodynamical sense, is it to discuss the exact geometry since we are in a situation described by polytopic-type bonds and geometries[9]. Therefore, in the study of clusters of water we expect the existence of many configurations with nearly the same energy, not only at finite temperatures, but also near zero temperature. As a consequence, in the study of the structure of water complexes, the emphasis should be on the probability distribution of the various configurations. We wish to point out that this situation occurring for the water complexes is not unique: since it has long been recognized that in the study of the structure of polymeric chains, the chemically significant problem is not the exact determination of the energetically lowest possible conformation, but rather the study of the probability distribution of all those conformations that are within an energy range statistically compatible with the temperature of the system in consideration[10]. We further note that the energy barriers in polymeric chains, the main factor controlling the distribution of conformations, are in the same energy range as the water-water binding energies (namely of the order of 3 to 6 kcal/mole).

The basic technique found useful in the conformational study of polymers can be adapted in the study of the conformations of the water complexes (or of ion-water complexes), namely an analytical potential of relatively simple form is constructed and iso-energy contour maps are obtained to aid in the selection of the most probable conformations.

Previous theoretical studies on the structure of water complexes[11,12] have, however, followed the traditional approach commonly used for the quantum chemical characterization of small molecules in the gaseous phase, where one attempts to determine the energetically most stable single geometry for the molecule. We are of the opinion that this commonly adopted approach might not be sufficiently meaningful since: (a) a number of competing structures will be neglected and (b) likely it will be practically impossible to obtain the lowest configuration since there are too many degrees of freedom to analyze, and presently a full search of the potential surface is too expensive for direct quantum mechanical computations.

Therefore, in this section we present a study of the water-water complexes[13] where use is made of a simple analytical potential obtained from direct computations in the Hartree-Fock approximation; with such a potential, assuming pair-wise additivity, a full search of the potential surface for the $(H_2O)_n$ complexes is performed. For the complexes $(H_2O)_3$ and $(H_2O)_4$, direct Hartree-Fock computations are presented to check the validity of the pair-wise addivity assumption. Previously we have reported the iso-energy contour maps for the Hartree-Fock interaction of two molecules of water, referred below as molecule A and molecule B, subjected to two geometrical constraints.

First, the geometry of each molecule was kept rigid (and chosen to be the known experimental geometry at equilibrium internuclear distances).

Second, both molecular planes were constrained to be perpendicular to each other, with the oxygen of the molecule B lying in the plane of the molecule A. This contour map can be constructed using the available analytical fit to the Hartree-Fock sampling of the interaction potential for A and B, and considering three variable geometrical parameters. If we fix rigidly the molecule A so that the oxygen atom is at the cartesian origin and the O-H bonds of A are in the x-y plane, then the variables are the x and y coordinates of the oxygen for B, and (for each pair of value of the variables) the rotational angle around an axis perpendicular to the molecular plane of A and passing through the oxygen nucleus of B. The number of configurations to be considered is high, in the order of several thousands, even with the assumptions implied in the above mentioned constraints.

We shall now relax some of the above constraints. For each position of the molecule B, its hydrogen is free to be either in or out of the plane. In other words, the rotational axis, passing through the oxygen nucleus of the molecule B, is not constrained to be perpendicular to the molecular plane of A, but can assume any direction. As known, any rotation can be obtained by performing three successive rotations with pre-established order and sense (for example, one can use Eulerian angles). The sampling of the interaction potential needed to obtain iso-energetic maps is now sharply increased, since there are no longer three geometrical variables, but five (the x and y coordinates of the oxygen atom of the molecule B and the three rotational angles).

This process can be repeated with the oxygen atom of the molecule B no longer constrained to be in the molecular plane defined by the molecule A, but free to be in any plane parallel to the one defined by the molecule A. This way only one constraint remains: the molecule is not allowed to vibrate.It is not a trivial task to find a practical and satisfactory solution to the problem of displaying the enormous number of configurations analyzed. In Fig. 5 we present a graphical solution that takes advantage of two computer programs kindly made available to us by D.A. Schreiber[14] and by C.K. Johnson[15]. For each oxygen-oxygen distance R_{0-0}, only the configuration corresponding to the lowest energy for the three angular rotations is reported. A master coordinate axis x, y, z is located at the oxygen nucleus of the fixed

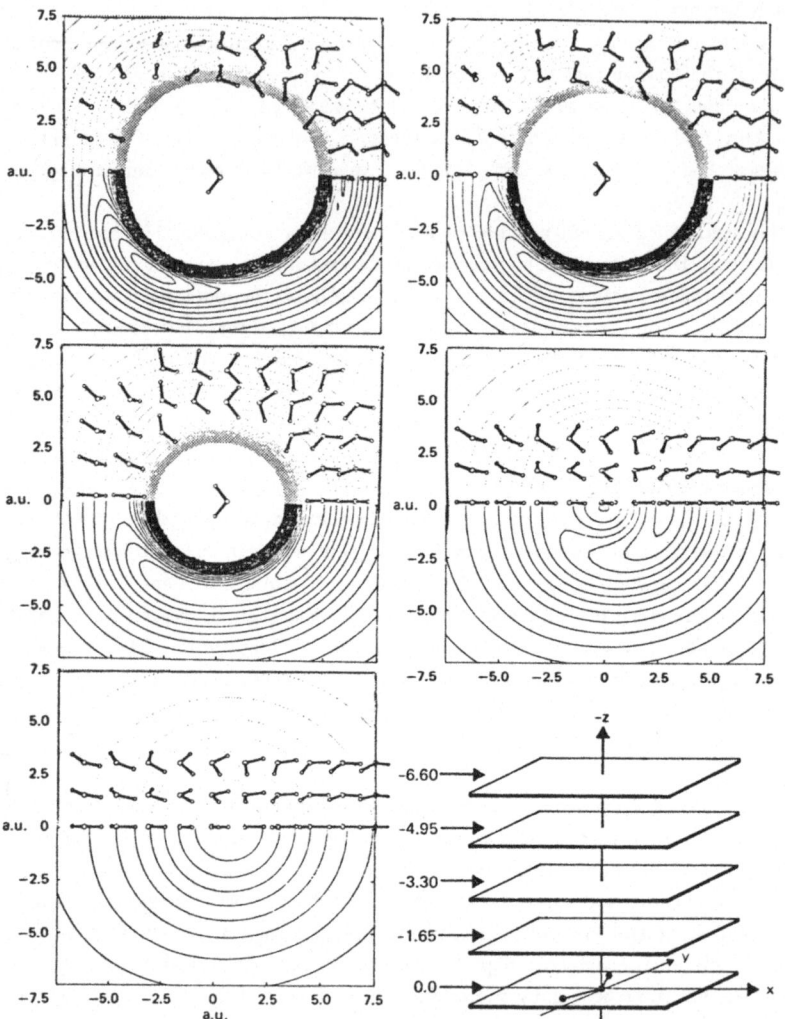

Figure 5 Minimal energy contour maps for the interaction of two molecules of water. One molecule is kept fixed at the origin of the coordinate system (see bottom right insert). The second molecule has its oxygen nucleus constrained parallel to the xy plane at z = 0.0 a.u. (top left insert), at z = 1.65 a.u. (top right insert), at ż = 3.3 a.u. (middle left insert), at z = 4.95 a.u. (middle right insert), at z = 6.6 a.u. (bottom left insert). The contour interval is 0.0005 a.u. The lowest contour (minimum) is at –4.4 kcal/mole with the exception of the bottom left insert, where the lowest contour is at –2.8 kcal/mole. The orientations of the second molecule are energy optimized and some of these orientations are given in a perspective view with the O–H distances being given only in half of their actual length and the viewer being perpendicular above the planes.

molecule, A, lying in the xy-plane, the x axis being, for example, oriented as to coincide with the C_2 rotational axis of molecule A (a water molecule has C_{2v} symmetry at the equilibrium configuration of the ground state). The oxygen atom of the molecule B is constrained to be either in the plane of xy (z=0.0) or in a number of planes parallel to the xy-plane. The planes we have considered are defined by z=0.00 a.u., -1.65 a.u., -3.30 a.u., -4.95 a.u., -6.6 a.u. The iso-energy contours in each plane have intervals of 0.0005 a.u. (about 0.3 kcal/mole). A computer program computes for a grid of points in each plane the energy of the A-B system, using the analytical fit to the Hartree-Fock potential, optimizes the molecular orientation based on a minimum energy criterion, and then transmits these data to the contour program[14]. The angular orientation of the hydrogen atoms (or of the O-H bonds) for the molecule B, corresponding to the lowest energy for the chosen R_{O-O} distance, is graphically given in Fig. 5 by presenting a three dimensional schematic representation of the B molecule. The O-H bond is not given in scale, but the O-H distance is represented by exactly one half of the correct one in order to display the data in a comprehensible way.

The display of the B molecules is restricted to a sufficient number so as to convey a visual "feeling" of the way the molecule B orients itself in space when it moves from point to point in a given plane. The main information we wish to convey is of qualitative nature (the quantitative information is in the analytical expression for the Hartree-Fock potential). From the information displayed in Fig. 5, a cross section of the energy surface, one can obtain a good representation of the surface. In the following we shall briefly comment on the cross section presented in this section. First, we have a repulsive region and an attractive region. The repulsive region can be identified very simply in the -0.0 a.u., in the -1.65 a.u. and in the -3.3 a.u. planes, since here the repulsion is very rapidly increasing when one brings the molecule B close to A: the contours are very near to one another, forming a ring, whose last contour represents a cut of 3.4 kcal/mole. We have not displayed more repulsive contours, since they are not too important for the study of systems near room temperature.

The attractive regions of the potentials are more important: the three minima of the 0.0 a.u. plane, corresponding to the open for of the dimer with nearly linear hydrogen bonds, are still present in the -1.65 a.u. plane, but not all of them appear in the -3.3 a.u. plane. The remaining minimum moves towards the x=y=0 position in the -4.95 a.u. plane and is nearly at this position in the -6.6 a.u. plane. It is noted that all the contours of the -6.6 a.u. plane are attractive. These are the main energetic features emerging from the Hartree-Fock potential: Fig. 5 contains a number of important details that will be discussed later. In Fig. 5 the energy of the lowest contours is -4.4 kcal/mole (with the exception of the lowest contour in the -6.6 a.u. plane, where the energy is -2.8 kcal/mole).

The orientation of the water B, relative to the fixed molecule of water A, is now briefly considered. Let us fix our attention on a molecule B on the C_{2v} axis, to the left of the molecule A, with the oxygen at about x=5.5 a.u., y=0.0 a.u. In the 0.0 a.u. plane, we can see only one O-H bond (looking from above the plane in the -z direction) since the second one is masked and exactly below. If we move this molecule along the -z axis (from the 0.0 a.u. plane to the -6.6 a.u. plane) and let the hydrogen find the energetically optimal configuration relative to the molecule A, we can learn from Fig. 5 that the molecule slowly rotates around an axis passing through the oxygen and parallel to the y axis; in the plane 0.0 a.u. the hydrogen points away from the A molecule, in the plane -6.6 a.u. the hydrogen points to the 0.0 a.u. plane, and one to the oxygen of the A molecule (the molecule A is not shown in the -4.95 a.u. and -6.6 a.u. planes).

Let us now consider in more detail the -3.30 a.u. plane. In Fig. 5 there is only one minimum; but one can see a rather extended region enclosed between the third and fourth lowest contour. In this flat region there is a very shallow minimum that can be obtained, for example, by selecting a smaller contour interval than the one of 0.0005 a.u. used for Fig. 5. To this shallow minimum corresponds a second form for a water dimer, designated as "closed form", containing two hydrogen bonds[13,16]. The closed form of the dimer is of interest as the simplest prototype of cyclic polymers of water (closed rings), whereas the open form is the prototype of the open chain polymers of water.

Finally it is noted that previous studies on the dimer like those of Marokuma and Pederson[3], Kollman and Allen[17], Del Bene and Pople[3] (all done in the SCF-LCAO-MO approximation) are not sufficiently accurate so as to be of little help in this problem. Other computations by Hankins, Moskowitz and Stillinger[8], by Diercksen[7], Lentz and Scheraga[12], have not scanned the potential space sufficiently as to be in a position to study the closed dimer structure.

Hankins, Moskowitz and Stillinger[11] have reported a Hartree-Fock study (with a sufficiently large basis set as to be near the Hartree-Fock limit) on the water trimer, and have concluded that triplets of water molecules with sequential hydrogen bonds stabilize significantly the condensed phase which incorporate such structures. A total of eight configurations have been analyzed by these authors. Since these authors have considered structures related to the cubic and the hexagonal forms of ice, from this work it is not easy to extract a general conclusion for the stable form of the water trimer.

Lentz and Scheraga[12] have recently examined the structure of the water trimer and the water tetramer with the aim to compare the binding energy of the cyclic and open form of these complexes. Their conclusion is that the cyclic trimer is less stable than the non-cyclic one, in contradiction with the results obtained by Del Bene and Pople[19]. It is noted that Lentz and Scheraga's computations have been performed in the Hartree-Fock model with adequately large basis sets, whereas Del Bene and Pople's basis set is sufficiently truncated as to give up to 2 kcal/mole error in the stabilization energy of small complexes, like dimers[3].

As remarked in the introduction to this chapter, and as evidenced by the two works summarized above, a full optimization of the geometry for a system containing several energy minima is a complex task. Clementi et al.[13] have chosen to use the analytical fit to the Hartree-Fock sampling (a-f-H-F) for the $(H_2O)_2$ complex to determine the optimal configuration on an energy criterium for the $(H_2O)_n$ (n=3,...,8) complexes. This might constitute a severe assumption, which however has been tested (see below) by considering the magnitude of the three and four body effects obtained by computations on the $(H_2O)_3$ complex and computations on the $(H_2O_4$ complex. For the accuracy required in an approximative structure determination of the $(H_2O)_n$ complexes, the previous additivity assumption appears satisfactory as shown below.

For the $(H_2O)_3$ complex, the best configurations optimization yielded a geometry given in Fig. 6. Thousands of possible configurations have been generated with the a-f-HF, translating and rotating the three molecules in many possible ways. The resolution of this search depends on the magnitudes of the steps between two successive geometries. The optimization method used a modified version of a Simplex computer program kindly supplied by J.P. Chandler[18]. Examining Fig. 6, we learn that the most stable geometry obtained has nearly C_3 symmetry, a result in agreement with Del Bene and Pople[19]; however, the dimension of the ring is significantly larger: our values for the oxygen-oxygen distances (see Fig. 6 for notation) are $R(O_1,O_2)$ =2.96A, $R(O_2,O_3)$=2.96A, $R(O_1,O_3)$ =2.97A to be compared with Del Bene and Pople's distance of $R(O_1-O_2)$=$R(O_2-O_2)$ =$R(O_1-O_3)$=2.56A[18].

The hydrogen bond angles in our computation are bent outwards by about 22°. The energy of stabilization for this structure obtained with the a-f-HF is 0.0196 a.u. (12.3 kcal/mole).

Using the same technique previously described for obtaining the best (most stable) configuration of the water trimer, the water tetramer was found to have the structure given in Fig. 6. The constraint imposed on the tetramer was to leave all oxygen atoms lying on the perimeter of a circle.

The geometrical characterization of this tetramer is as follows: $O(1)-O(2)$ =2.95A, $O(1)-O(3)$ = 4.17A, $O(1)-O(4)$ = 2.95A, $O(2)-O(3)$ = 2.95A, $O(2)-O(4)$ = 4.15A, $O(3)-O(4)$ = 2.95A; the angular characterization is given in reference[13].

As an introductory analysis to larger complexes of water, we have analyzed[13] a very restricted class of polymers of water, which have been discussed amply in literature but not studied in a quantitative way. We feel that the use of the additivity approximation will yield reliable indication on the best structure and

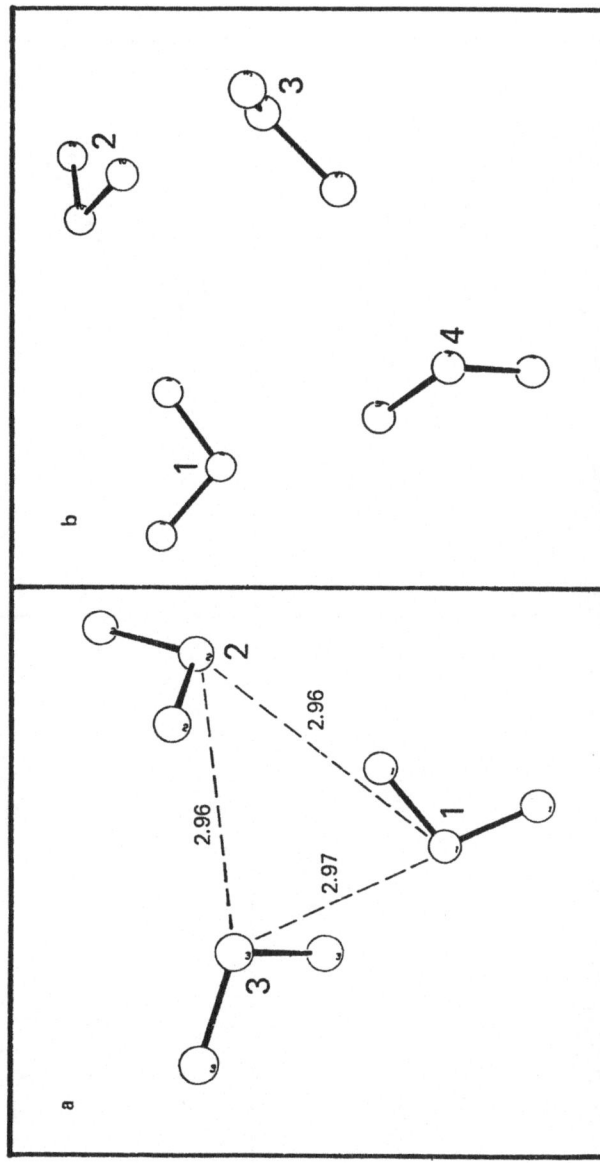

Figure 6 Most stable comfigurations for the trimer (insert a)) and the tetramer (insert b)) of water at T = 0°K (see text for additional explanations about the geometry).

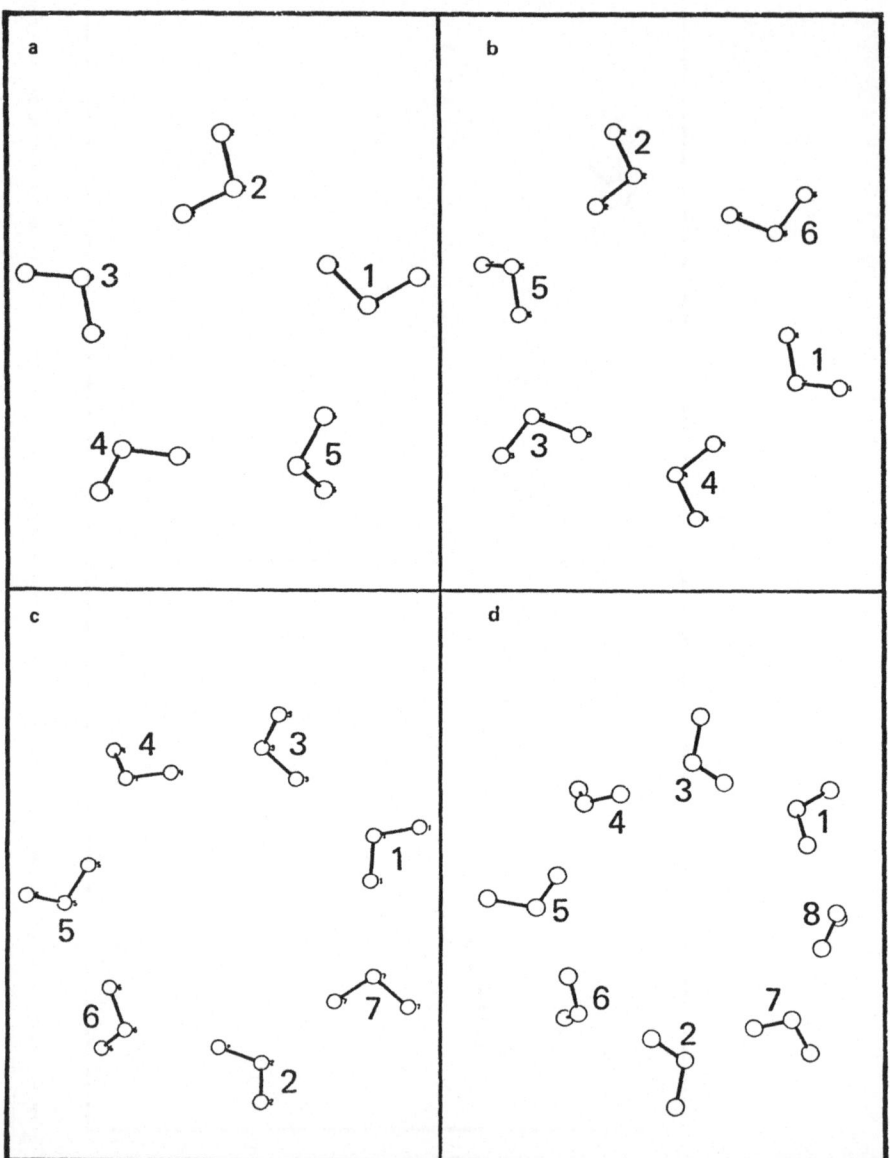

Figure 7. Most stable configurations for $(H_2O)_5$, $(H_2O)_6$, $(H_2O)_7$ and $(H_2O)_8$ obtained under the constraint that the oxygen nuclei have to lie on the periphery of a ring. The hydrogens are free to assume any orientation.

stabilization energy. With the constraint that the oxygen atoms should be on the periphery of a circle, that means we require all oxygens to be in the same plane, and with no constraint for the hydrogen orientation, we have searched for the configuration of minimum energy for $(H_2O)_5$, $(H_2O)_6$, $(H_2O)_7$, and $(H_2O)_8$. The resulting structures are displayed in Fig. 7. For all these rings one hydrogen for a given molecule is hydrogen-bonded and the second one tends to minimize the hydrogen-hydrogen repulsion. We shall not discuss at length these structures, but only some that characterize the result given in Fig. 7.

For the pentamer, $(H_2O)_5$ the geometrical characterization is: $O(1)-O(2) = 2.97A$, $O(1)-O(3) = 4.80A$, $O(1)-O(4) = 4.79A$, $O(1)-O(5) = 2.96A$, $O(2)-O(3) = 2.97A$, $O(2)-O(3)$ $4.80A$, $O(3)-O(4) = 4.80A$, $O(3)-O(4) = 2.97A$, $O(3)-O(5) = 4.80A$, $O(4)-O(5) = 2.95A$.

The hydrogens that are hydrogen bonded are characterized by the angles: $\alpha = 4° \pm 2°$, $\beta = 3° \pm 1°$, $\gamma = 1° \pm 1°$. One can notice that the hydrogen bonds are somewhat more straight than in the tetramer case: this is expected since there is less strain the larger the ring.

For the hexamers, $(H_2O)_6$ the geometrical characterization is as follows: $O(1)-O(2) = 5.13A$, $O(1)-O(3) = 5.13A$, $O(1)-O(4) = 2.96A$, $O(1)-O(5) = 5.92A$, $O(1)-O(6) = 2.96A$, $O(2)-O(3) = 5.12A$, $O(2)-O(4) = 5.92A$, $O(2)-O(5) = 2.96A$, $O(2)-O(6) = 2.96A$, $O(3)-O(4) = 2.96A$, $O(3)-O(5) = 2.96A$, $O(3)-O(6) = 5.92A$, $O(4)-O(5) = 5.12A$, $O(4)-O(6) = 5.13A$, $O(5)-O(6) = 5.12A$. The value of the angle is $3 \pm 1°$ with exception of $\alpha (H_5O_5O_3) = 7°$; β is equal to $2° \pm 1°$; and γ is equal to $3° \pm 1°$ with exception of $\gamma (H_5O_5H_5^+) = 7°$.

For the heptamer, $(H_2O)_7$ the following geometrical characterization is given: $O(1)-O(2) = O(1)-O(4) = O(2)-O(5) = O(3)-O(5) = O(3)-O(7) = O(4)-O(6) = O(6)-O(7) = 5.32 \pm 0.01A$; $O(1)-O(3) = O(1)-O(7) = O(2)-O(6) = O(2)-O(7) = O(3)-O(4) = O(4)-O(5) = O(5)-O(6) = 2.95 \pm 0.01A$, and $O(1)-O(5) = O(1)-O(6) = O(3)-O(6) = O(4)-O(7) = O(5)-O(7) = 6.65 \pm 0.01A$. The values of the angles are $\beta = 6° \pm 1°$. $\beta = 5° \pm 1°$ and $\gamma = 2° \pm 1°$ with exception of $\gamma (H_3O_3H_3) = 5°$.

The octamer, $(H_2O)_8$ has the following geometrical characterization: the nearest neighbor has $(O_i)-(O_j) = 2.96$, $\alpha = 8° \pm 1°$, $\beta = 8° \pm 2°$ and $\gamma = 3° \pm 1°$. The next nearest neighbor has a $(O_i)-(O_j)$ distance of $5.46A$, the next nearest neighbors have $(O_i)-(O_j) = 7.13A$.

In a large number of calculations we relaxed the symmetry constraints imposed on the positions of the oxygen nuclei. At first we required the oxygen nuclei to lie on the periphery of a sphere, thus admitting non-planar ring structures. The resulting structures differ not too much in energy and geometry from the planar rings, however the larger clusters show puckered non-planar rings. Relaxing all constraints we obtain for $(H_2O)_4$ and $(H_2O)_5$ ring structures, but for the larger clusters we obtain for a given cluster size various irregular structures with multiple branching of H-bonds on one molecule (corresponding to open H-bonds) and considerably different geometries but quite similar energies. For instance, in Fig. 8 three forms for the $(H_2O)_8$ complex are given. The corresponding stabilization energies starting from the right are 47.30 kcal/mole, 46.95 kcal/mole, and 46.76 kcal/mole. This proves the necessity to study the statistical distribution of configurations and thus reduces the importance of individual geometries.

In Fig. 9 we display the energy stabilization and the stabilization energy per molecule as a function of the polymer size n. For the planar rings, the spherical rings, and the irregular structures the stabilization energies are quite similar, the spread is not large, but within this spread a large number of different structures are existing, increasingly larger the larger the polymer size. The binding energy per molecule is increasing with increasing cluster size, this indicates the stability of the larger clusters against dissociation in any combination of smaller ones, at least at T=0°K. The binding energy per molecule is much smaller than the corresponding value of a large system of water molecules (for instance at T=273°K Monte Carlo calculations with the a-f-H-F potential yield a value of 7.1 kcal/mole). The relatively small stabilization energy per molecule and the fact that the open structures are more stable for larger clusters indicates that it does not seem to be possible to describe the energetic properties of water as resulting from an ideal mixture of small regular clusters[19]. This conclusion would not be changed even when we take into account

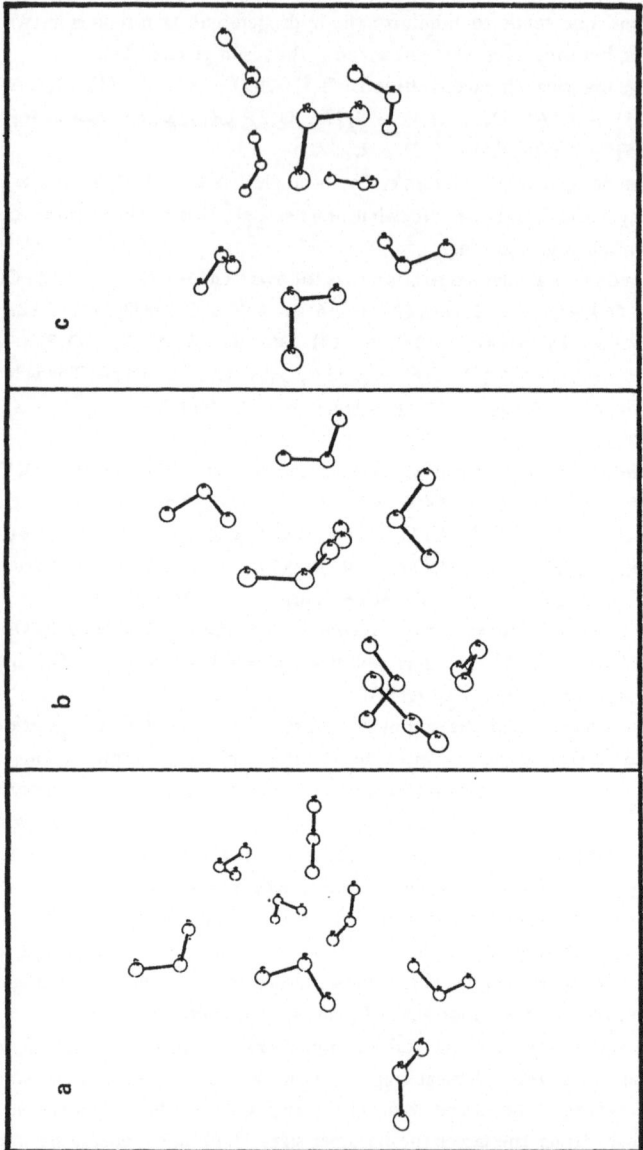

Figure 8   Different irregular structures for the energy optimized $(H_2O)_8$ complex, with no constraints imposed.   Starting from the left the corresponding stabilization energies are 47.30 kcal/mole, 46.95 kcal/mole and 46.76 kcal/mole.

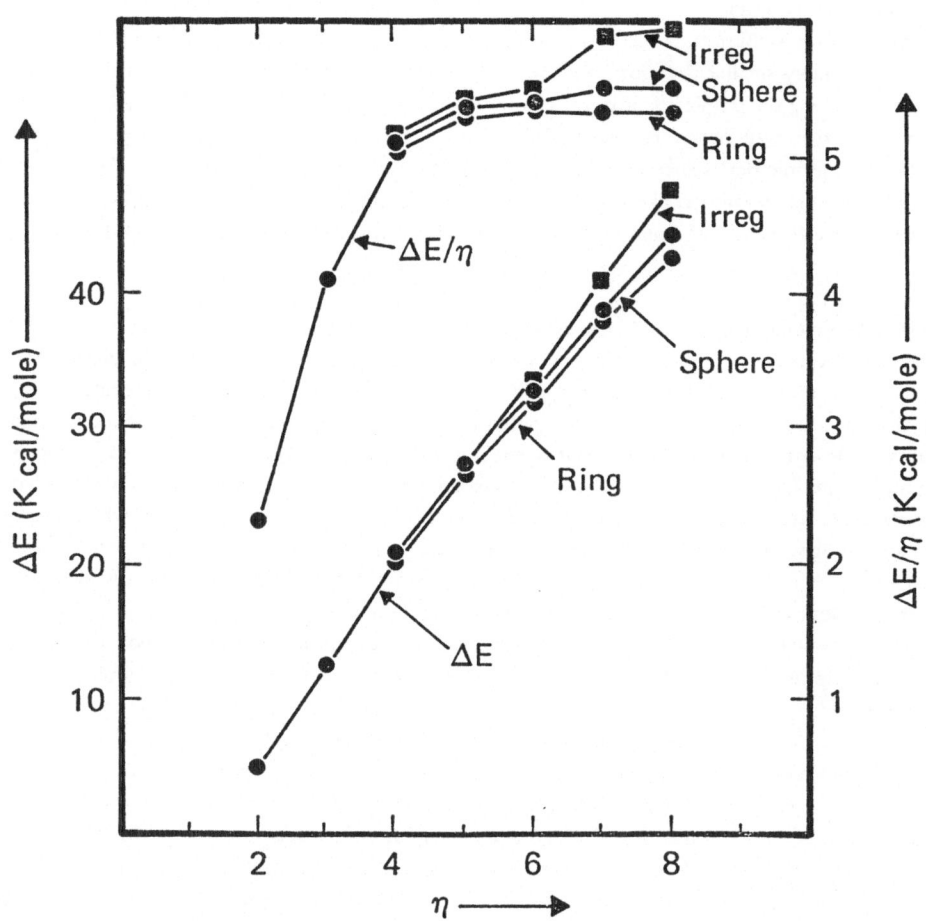

Figure 9 Stabilization energies for the $(H_2O)_n$ clusters as a function of the cluster size n. The stabilization energies have been obtained by imposing three different constraints. In the first case the oxygen nuclei have been required to be on the periphery of a planar ring (designated by "ring" in Figure 10). In the second the oxygen nuclei have been required to be on the periphery of a sphere (designated by "sphere" in Figure 10). In the third there have been no constraints imposed (designated as "irreg" in Figure 10).

three-body effects for the smaller clusters. This is somewhat in contradiction with the results obtained by Pople and Del Bene[19], which would favour a mixture model, with the trimers being the dominant structure[12]. As noted before, Pople and Del Bene's result is probably caused by the use of a too much truncated basis set. As pointed out by Ben Naim[20] any liquid can be formally regarded as a mixture of quasicomponents, for instance clusters, but according to our results this does not seem to be very convenient for liquid water, since the mixture would be strongly non-ideal and one would probably need a large amount of different clusters, differing both in size and geometry. But again, as noted before, one has to be very careful in transferring this conclusion to systems at finite temperature, since there entropy effects may be very important.

We concluded that the stabilization energy per molecule shows no special stability of an individual cluster and is also considerably smaller than for liquid water. This seems to be in contrast with mixture-models of liquid water, describing the properties of water as resulting from an ideal mixture of small regular polymers. Due to many energetically nearly equivalent but geometrically quite different structures for the larger clusters, the exact geometrical structure is less important. The probability distribution of different cluster structures at finite temperatures is likely a more physically meaningful parameter. In addition it has to be noted that entropy effects have an important influence on the relative stability of different clusters.

2.4 Monte Carlo Simulation of Liquid Water using an Hartree-Fock Potential

If a satisfactory account of the structure of aqueous systems is to be developed, calculations of the structure of water and of ionic solutions should be based on the same model of water-water interactions.

Although the work limited to an Hartree–Fock potential and referred to in the previous section provides a good qualitative description of small clusters of water, it is known that the Hartree-Fock interaction ignores certain contributions to the potential energy of the system, the correlation energy corrections. Correlation energy terms are present over the whole range of interaction. At very short distances, where the two molecules are strongly repelling, the Hartree-Fock representation is likely to be sufficiently accurate, as it is in the neon-neon interaction. Unfortunately, water molecules are not likely to be sufficiently close to the liquid phase at room temperature conditions for the strongly repulsive region to be important and some attempt must be made to include the correlation terms. As discussed in the first part, it is possible to obtain an accurate description of the effect of including correlations terms by examining a number of semi-empirical forms[21].

It is known that at large distances the dominant contribution to the correlation energy arises from the induced dipole-induced dipole interaction, a term that is proportional to the inverse sixth power of the distance between the centers of the molecules. However, at shorter distances other contributions are more significant and a more complicated form must be assumed for the correlation energy. In this section a semi-empirical form is used for the correlation energy at short and intermediate distances and the effect of including this with and without the long range form is examined. It is shown that although there are significant differences between the several combinations proposed, they all retain the property of having broad minima in the hydrogen bonding configurations. Examining the liquid state calculations it is shown that the theoretical results are consistent with the changes in the depths of the potential minima and that there are a number of problems to be solved before the experimental scattering data will be acceptable.

The semi-empirical form for the correlation energy is based on an expression suggested by Wigner[22] from work on electron gas and discussed in chapter 1. As the intermolecular potential is defined as the difference between the total energy of the molecules at a particular distance and relative orientation and the energies of the isolated molecules the required correction term will be of the form:

$$\Delta E_c = E_c(d) - 2E_c(M) \tag{1}$$

Δ is the "molecular-extra correlation energy" of the dimer and $E_c(M)$ is that of the isolated molecule.

In this section we discuss a semi-empirical form used to describe the correlation energy of the dimer at short and intermediate distances for a given configuration[21] namely:

$$\Delta E_c = \int [\, (a_1 \, \rho_d \, 4/3 \,) / (a_2 + \rho_d 1/3)^{-1} \ (2a_1 \, \rho_M \, 4/3 \,) / (a_2 + \rho_M \, 1/3^{-1} \,] \, d\rho \qquad (2)$$

In this equation, ρ_d is the Hartree-Fock density for the water dimer in a given configuration, ρ_M is the Hartree-Fock density for the isolated molecule, and the constant a_1 is chosen so that the exact correlation energy of the monomer is obtained. Equation (2) was used to calculate the correlation energy correction for a number of dimer configurations and separations. It was found that ΔE_c was relatively insensitive to the relative orientation of the dimers at a given separation and consequently the results were fitted to an expression

$$\Delta E_s \, (r) = - c \, \exp \, (-\delta r) \qquad (3)$$

where r is the oxygen-oxygen separation and the subscript "s" is used to emphasize the essentially short range nature of this correction term. The coefficients giving the best fit were $\delta=1.843A^{-1}$ and $c=64.64$ kcal/mole.

Since the Hartree-Fock Hamiltonian and the resulting densities take no account of the dispersion potential it is necessary to include such terms. The leading contribution at large distances is given by the induced dipole-induced dipole potential

$$E_D \, (r) = -C_6 \, / \, r^6 \qquad (4)$$

where we take r to be the oxygen-oxygen distance as before. Accurate values of the coefficient C_6 have not been calculated for the water dimer and so in this work two extreme values are used, that obtained from the Kirkwood-Müller relation[23] $C_6=1222.2$ kcal/mole$^{-1} A^6$, and that obtained from the London relation[23] $C_6=676.5$ kcal/mole$^{-1} A$. Although in principle the potential given in equation (4) should be angle dependent, the polarizability tensor of the water molecule is not significantly anisotropic and therefore orientational effects have not been considered[24].

In order to obtain an accurate impression of the importance of the correlation terms, a number of potentials were constructed. First the pure Hartree-Fock potential (HF) was considered and then the Kirwood-Müller (HF+K) and London (HF+L) dispersion terms were added. To examine the influence of the short ranged correlation correction two further potentials were constructed, HF+W+K(s) and HF+W+L(s). In both these potentials the long-range terms were modified by introducing a switching function[25] such that at r≤3A there was no contribution from the dipole-dipole dispersion term and at r≥ 4A this term is unaltered. Finally, a potential consisting of the Hartree-Fock potential with the Wigner function above (HF+W) was examined.

The intermolecular potentials for the water dimer were compared in two ways, by examining the dependence on distance for fixed relative orientations and by examining energy contours around a fixed molecule. As there are many relative orientations of the dimer that can be considered some choice had to be made between them. It was decided to examine in detail the dependence on distance of the open dimer configuration reported in previous work[13]. This relative orientation includes the most stable geometry for the dimer determined on the basis of Hartree-Fock calculations. Briefly, the x-axis contains the two oxygen atoms and one hydrogen atom lying between the oxygens - a typical hydrogen bonding orientation. To obtain the distance dependence of the potentials the y and z coordinates of both molecules were fixed and the x coordinate of one molecule varied. The distance was defined as the oxygen-oxygen distance. Fig. 10 compares the potential HF with that termed HF+L, Fig. 11 compares HF with HF+K, Fig. 12 compares HF with both HF+W and HF+W+L(s). A number of interesting observations can be made.

To begin with, consider the equilibrium distances predicted by the different potentials. The potential HF has a minimum at a distance of 2.97±0.03A, the error bounds being determined from Hartree-Fock calculations using several different basis sets. All the potentials containing corrections for the correlation energies shift the equilibrium positions towards smaller values of the intermolecular separation, the largest shift arising from the Kirkwood-Müller term (-0.1A), the smallest shifts being found when only the Wigner term is included (-0.2A). This observation is to be expected as all the correction terms are negative and the Kirkwood-Müller term is the strongest. Observations on a large body of Hartree-Fock calculations suggest that the position of the minimum in the HF potential is very close to the correct equilibrium value. Consequently, it is likely that both the HF+K and HF+L potentials overestimate the shift in the minimum and that the three potentials including the Wigner term are accurate.

Next, consider the value of the energy at the minimum as this represents the binding energy or hydrogen bond energy for the system. The potentials have minima of -4.6 kcal/mole (HF), -5.6 kcal/mole (HF+L), -6.6 kcal/mole (HF+K), and -4.9 kcal/mole. HF plus London switched plus Wigner. The potentials containing the Wigner correction all have the same minima by construction. As the experimental binding energy[18] is about -6± 3kcal/mole and the energy spread in the present potentials is from -4.6 kcal/mole to

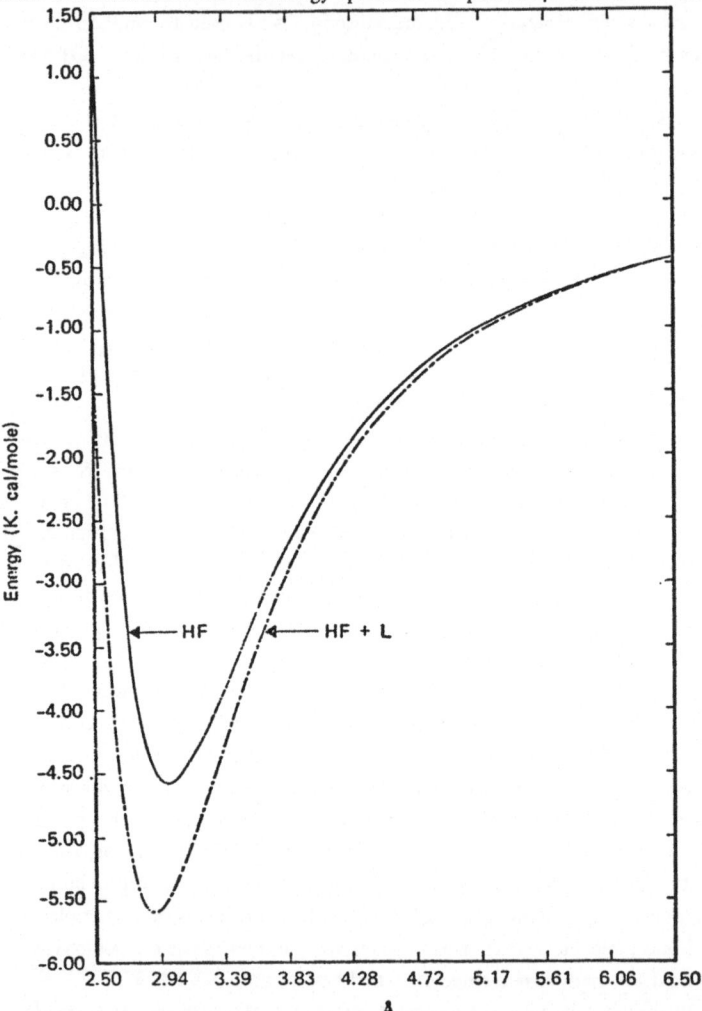

Figure 10 Hartree–Fock (upper curve) and Hartree–Fock corrected by the London term (lower curve) potentials for the water–water interaction. The energy scale is in Kcal/mole and the oxygen–oxygen distance scale is in Å. The position of the hydrogens relative to the oxygen in each molecule of water, and the relative orientation of the two molecules of water has been kept constant for all the oxygen–oxygen separations.

-6.6 kcal/mole it is clearly not feasible to distinguish between the various models by comparing with experiment. An unambiguous decision on the current binding energy cannot be made before a full quantum mechanical calculation including correlation effects is presented. However, it is most probable that the correct value lies within the range of potentials used here. Consequently, we can be confident that the conclusions to be reached on the importance of the correlation corrections are meaningful, as we shall prove in the next section.

The Monte Carlo program used to simulate the behavious of liquid water has been described elsewhere[23]. Briefly, a large number of configurations (between 5×10^5 and 10^6 in this work) of a system of N molecules enclosed in a cubic box with periodic boundary conditions are generated in such a way that the probability of a given configuration being observed is proportional to the Boltzmann weighting for that

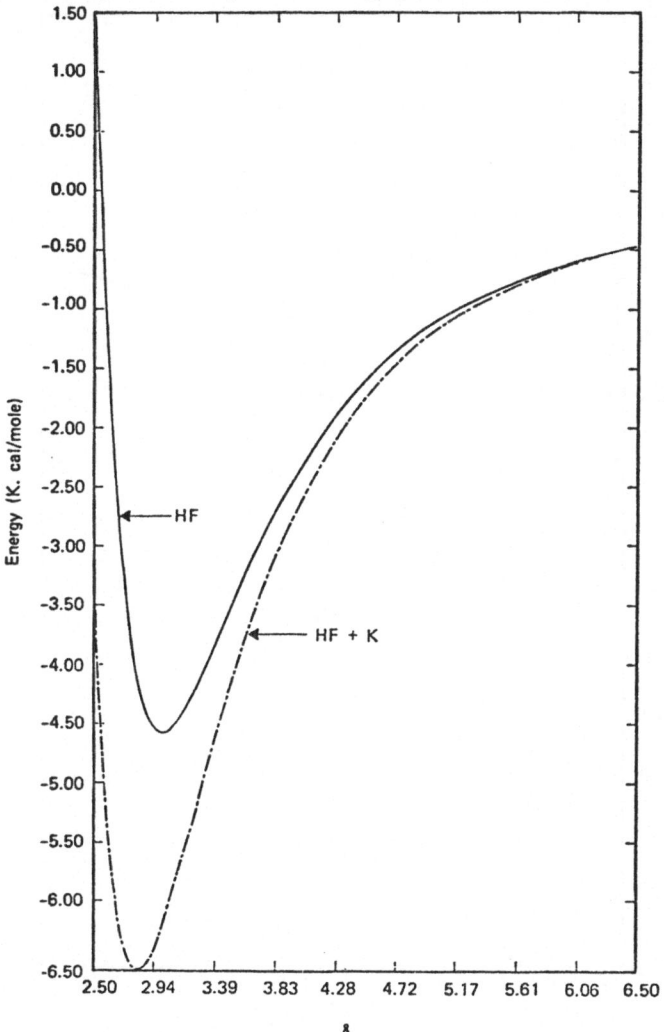

Figure 11 Hartree–Fock (upper curve) and the Hartree–Fock corrected by the Kirkwood–Müller term (lower curve) for the water–water system. (See **Fig. 1** for additional comments.)

configuration. Averages over certain properties of each configuration enable accurate estimates to be made of thermodynamic and structural data for the system (for example averaging the total potential energy of each configuration enabls accurate estimates of the thermodynamic internal energy). In this work the internal energy and specific heat of systems of either 64 or 125 particles were calculated together with certain structural properties while keeping the geometry of the H_2O molecule rigid. The radial distribution function of a system measures the probability of finding a particle at a given distance from a central particle and for water three such functions are of interest. Oxygen-oxygen correlations are measured by the radial distribution function $g_{OO}(r)$, oxygen-hydrogen correlations by $g_{OH}(r)$ and hydrogen-hydrogen correlations by $g_{HH}(r)$. These functions have been previously reported for the HF potential only. Given that the functions $g(r)$ measure the probability of finding a particle at a distance r from a given particle it follows

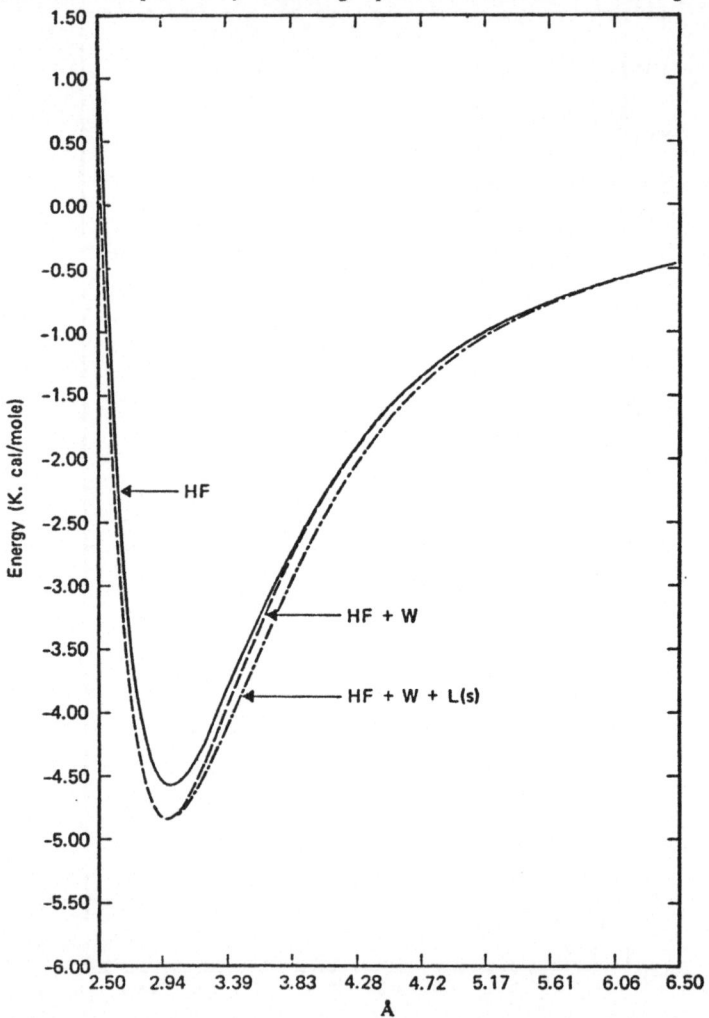

Figure 12 Hartree–Fock (HF, upper curve), Hartree–Fock corrected by the Wigner and London term (HF+W+L(s); the latter is switched off at the equilibrium separation) potentials for the water–water system (see Fig. 1 for additional comments).

that the number of particles in a spherical shell of radius r and thickness dr about a given particle is

$$\Gamma(r) = 4\pi\rho r^2 \, g(r) \tag{5}$$

where $\rho = N/V$ is the number density of the system. In this work Γ_{OO}, Γ_{OH} and Γ_{HH} will be considered. Finally, the numbers of particles inside a shell of radius r about a given particle are given

$$N(r) = \int_0^r \Gamma(r) \, dr \tag{6}$$

and again in this work N_{OO}, N_{OH} and N_{HH} are of interest. The Monte Carlo estimates for the several model potentials were performed at a temperature of 298°K at the experimental density of liquid water. Results for the distribution functions were compared with the experimental data of Narten[26], and it is worth noting at this point that the experimental results for the functions $g_{OH}(r)$ and $g_{HH}(r)$ are not independent[26] and also depend on a model of the liquid. Finally, it is worth noting that there was no significant difference between the distribution functions calculated using 64 and 125 particles. Results were reported for the HF, HF+L and HF+K potentials as these cover the likely range of the correlation energy correction.

Fig. 13 reports the functions $g_{OO}(r)$, $\Gamma_{OO}(r)$, and $N_{OO}(r)$ for the three potentials and compares them with experimental data. All three potentials are different from the experimental result and in addition definite differences can be observed between the calculations. It is apparent that the distribution functions given by the theoretical potentials are less structured than is suggested by experiment and all three results have the second maximum shifted towards longer distances. This observation can be connected to the tendency towards spherical symmetry about the oxygen atom observed in the energy contour maps. Differences between the theoretical calculations are noticeable in three areas. As the strength of the induced dipole-induced dipole interaction increases the position of the first peak shifts and its height increases. The increase in height is to be expected as the deeper the potential at its minimum, the stronger the attractive forces in that region. Similarly we can expect the peak to move out as the stronger attractive forces should also reduce the pressure of the system. Examining the function $\Gamma_{OO}(r)$ confirms the observations made above and in particular it is apparent that the experimental results show more structure in the neighborhood of the second peak. The number of oxygen atoms within a distance r of a given oxygen atom, $N_{OO}(r)$, varies between potentials and between potentials and experiment. If the nearest neighbors shell is defined as containing all the particles in a sphere at radius 3.3A, the position of the first minimum in the experimental $g_{OO}(r)$, we see that the HF and the HF+L results are in good agreement with experiment in predicting about 4.5 molecules. The HF+K potential has a deeper minimum in the potential and predicts rather more neighbors, around five.

Restricting the discussion to the oxygen-oxygen distribution and particularly to the $\Gamma_{OO}(r)$ data, it can be seen that the theoretical potentials give a nearly accurate qualitative account of the structure of liquid water. The inclusion of exact correlation energy corrections (almost certainly encompassed by the HF and HF+K potentials) will not bring the theoretical potential into agreement with experiment and it is apparent that the importance of a careful analysis of the correlation correction and of three-body (but unlikely of higher order terms) must be examined.

2.5 The Structure of Liquid Water using an Accurate Potential

Recently, Diercksen. Kraemer, and Ross[27] reported extensive CI calculations for the water dimer. Although their calculations are limited to a relatively small portion of the potential surface in the vicinity of the equilibrium geometry, comparison with their results and with the dispersion correction results calculated by the perturbation technique[28] provides a valuable check on the accuracy of our correlation energy results.

However, a potential sufficiently extended in space, and not restricted around the minimum energy of the

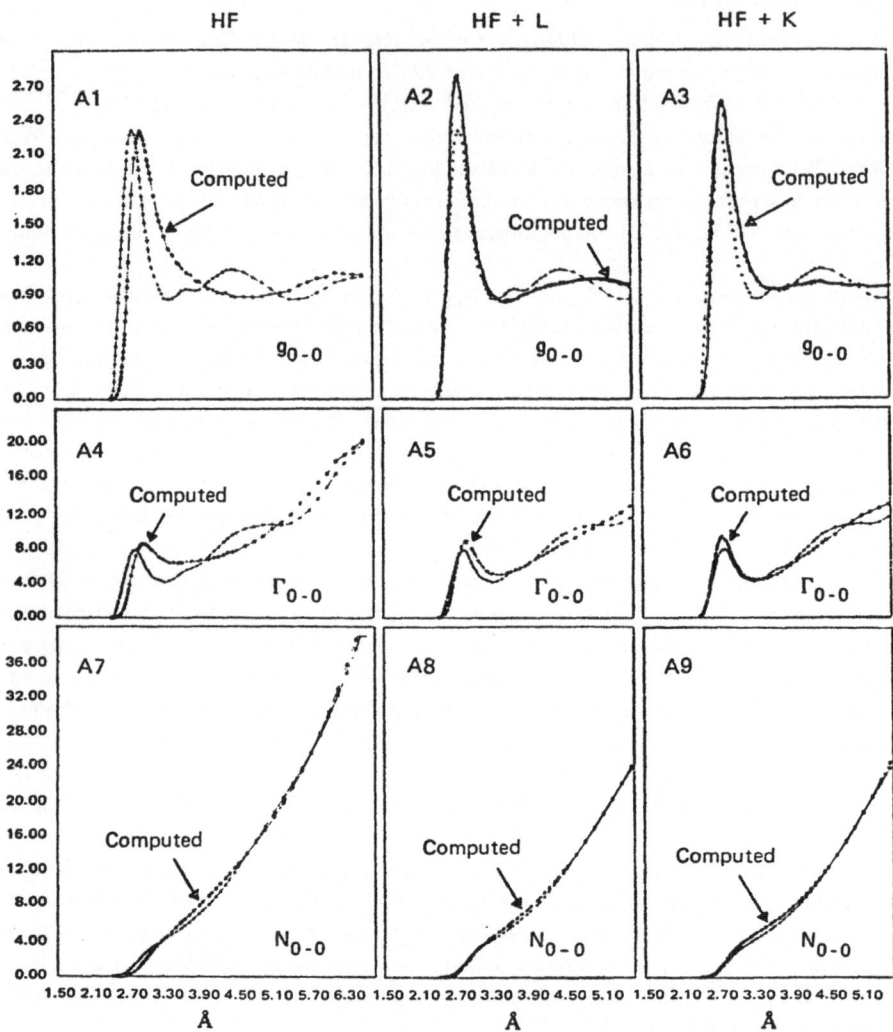

Figure 13 Oxygen–oxygen distribution functions for the HF, HF+L and HF+K potentials. From top to bottom are found $g_{O-O}(r)$, $\Gamma_{O-H}(r)$ and $N_{O-O}(r)$. The distances are in Å.

dimer, is now available[29], as below described. The electronic wavefunctions and energies for the ground state of the water dimer have been calculated using the configuration-interaction method. The ground state wavefunction is expanded in a form

$$\Psi = C_{SCF} \, \Phi_{SCF} + \sum_i C_i \, \Phi_i{}^{INTER} + \sum_i C_i \Phi_i{}^{INTRA} \tag{7}$$

where C's are variationally determined coefficients, and Φ are orthonormal configuration state functions (CSF). Since we were interested in mapping a potential surface for the water dimer in any possible conformation, we treated the system without reference to point symmetry consideration. Thus, each Φ is linear combination of Slater determinants (SD) constrained only to be a singlet state function. The SD's are built from an orthonormal one-particle basis set of spacial orbitals, which in turn are expanded in terms of a set of elementary basis functions (EBF). In equation (7) the wavefunction is written as a sum of three classes of Φ's. The first class consists of the SCF state function, Φ_{SCF}, which is taken as the reference CSF in this work. The second and third classes consist of Φ^{INTER} and Φ^{INTRA}, which contribute mainly to either the inter- or intra-molecular correlation energy, respectively. Later in this section we shall give precise definitions of these classes of CSF and the details on how these CSF's are constructed.

The elementary basis set we have used consists of contracted Gaussian-type (CGTF) centered at each atom; the type, the orbital exponent and contraction coefficient for each CGFT are listed in reference[29]. This essentially the same as the one used in the previous SCF water dimer potential calculations, and consists of (11,7,1/4,3,1)-functions (which means 11 s-, 7 p-, and 1 d-type functions contracted to 4 s-, 3 p-, and 1 d-type functions) centered on each oxygen atom and (6,1/2,1)-functions on each hydrogen atom.

A SCF calculation with the elementary basis set just described yields 10 canonical SCF occupied orbitals and 48 virtual orbitals. In our CI calculations we used as the core and internal orbitals, a set of orbitals obtained by localizing these canonical SCF orbitals on each monomer following the method developed by Edmiston and Ruedenberg[30]. This transformation was performed in order to ensure a meaningful partitioning of the dimer correlation energy to inter- and intra- parts. In addition, this transformation provides a reasonable way of selecting a limited number of external orbitals to be used in constructing CSF's. We write the SCF state function as

$$\Phi_{SCF} = 1 \, \Phi^2 2 \, \Phi^2 3 \, \Phi^2 4 \, \Phi^2 5 \, \Phi^2 6 \, \Phi^2 7 \, \Phi^2 8 \, \Phi^2 9 \, \Phi^2 10\Phi^2 \tag{8}$$
$$\quad\; A \quad\;\; A \quad\; A \quad\; A \quad\; A \quad\; A \quad\; A \quad\; A \quad\; A \quad\; A$$

where the subscripts A and B designate the water monomer on which the orbitals are localized; $1\phi_A$ and $2\phi_B$ are essentially the 1s oxygen orbitals that we shall call the core orbitals. The remaining 8 localized orbitals will be designated as the internal orbitals.

With this order of the core and internal orbitals, we define two classes of CSF as follows:

$$\Phi^{INTER} \equiv \left\{ \Phi_{ij,\,\alpha}^{k\ell} \right\} \qquad i = 3, ..., 6; \quad j = 7, ..., 10; \quad k\ell = 11, ..., 58, \tag{9}$$

$$\Phi^{INTRA} \equiv \left\{ \Phi_{ij,\,\alpha}^{k\ell} \right\} \qquad i,j = 3, ..., 6, \text{ or } 7, ..., 10; \quad k\ell = 11, ,..., 58, \tag{10}$$

where $\{\ \}$ means a CSF with state α constructed by exciting a pair of electron from i and j internal orbitals into k and ℓ SCF virtual orbitals. All the possible states arising from an orbital configuration which interact with the ϕ_{SCF} are included; the number of CSF in Φ^{INTER} and Φ^{INTRA} are 36,864 and 37,057, respectively.

In order to obtain an overall description of the water-water potential function, it is necessary to cover a wide range of geometrical configurations of the water dimer.

In selecting geometrical configurations we use the following approach. First, we fix a water molecule in a coordinate frame with O_1 at the origin and with H_1 and H_2 in the xy plane with the x axis bisecting the HOH angle (see Fig. 14). We note here that the geometry for the water molecule is always kept constant with the experimental structural parameters of H_2O ($R_{OH}= 0.9572$ A and $<HOH= 104.52°$) [31] . We then

let the second water molecule approach the first from different directions, and these directions are used in this work to define different types of geometrical configurations. We have selected 10 directions (see Fig. 14). From the geometries of Fig. 15, the B curve is clearly a potential curve for the bifurcated dimer.

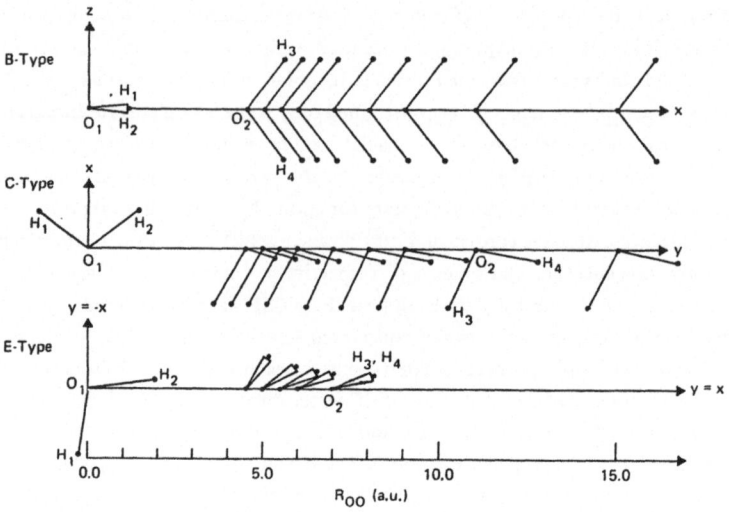

Figure 15a. Geometries of the H_2O dimer for various R_{OO}.
H_1 and H_2 are in the xy plane for all types.
B-type: H_3 and H_4 are in the xz plane.
C-type: H_3 and H_4 are off and opposite sides of the xy plane at small R_{OO}.
E-type: The xy plane is the symmetry plane of the second H_2O molecule.

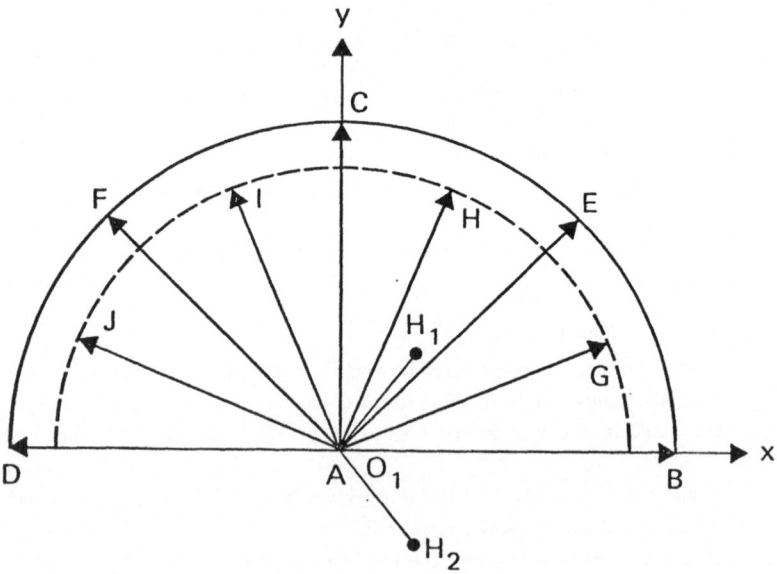

Figure 14 Geometry of the first H_2O molecule and 10 directions adopted to define geometrical configuration types of the H_2O dimer (see also Table 3).

Similarly, the E curve may be identified as that for the linear dimer, and its geometry at $R_{OO}= 5.5$ a.u. indicates it is very close to the optimum geometrical configuration for the water dimer. The geometry for the C curve is not exactly the symmetric cyclic structure, but it is close enough to be classified as a somewhat distorted cyclic structure in the vicinity of equilibrium. Since this region of the potential surface appears to be rather flat, evidenced by the closeness of the curves C, H and I, we shall take the C curve to represent the cyclic dimer.

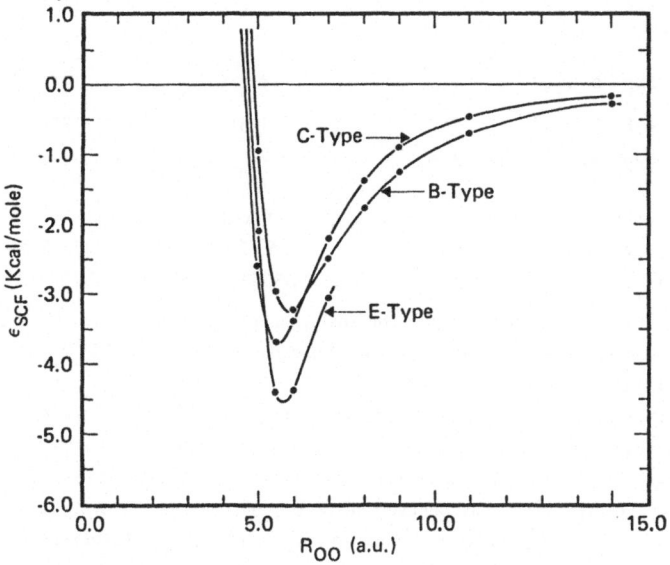

Figure15b. SCF H_2O dimer potential curves for B-, C-, and E-type geometrical configurations.

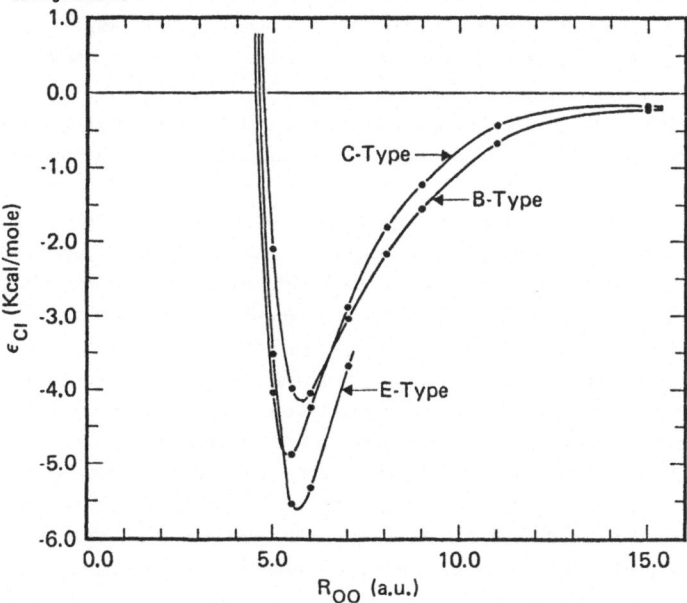

Figure15c. CI H_2O dimer potential curves for B-, C-, and E-type geometrical configurations.

Thus, the potential minima for the linear, cyclic, and bifurcated dimers are estimated to be respectively -5.6 (2.98 A), -4.9 (2.87 A), and -4.2 (3.01 A) kcal/mole (the corresponding O—O distances are given in parentheses).

The form of the potential function used is

$$
\epsilon = q^2 \left[\frac{1}{r_{13}} + \frac{1}{r_{14}} + \frac{1}{r_{23}} + \frac{1}{r_{24}} \right] + \frac{4q^2}{r_{78}} - 2q^2 \left[\frac{1}{r_{18}} + \frac{1}{r_{28}} + \frac{1}{r_{37}} + \frac{1}{r_{47}} \right]
$$

$$
+ \ a_1 \ \exp(-b_1 \ r_{56})
$$

$$
+ \ a_2 \ [\ \exp(-b_2 \ r_{13}) + \exp(-b_2 \ r_{14}) + \exp(-b_2 \ r_{23}) + \exp(-b_2 \ r_{24}) \ ]
$$

$$
+ \ a_3 \ [\ \exp(-b_3 \ r_{16}) + \exp(-b_3 \ r_{26}) + \exp(-b_3 \ r_{35}) + \exp(-b_3 \ r_{45}) \ ]
$$

$$
- \ a_4 \ [\ \exp(-b_4 \ r_{16}) + \exp(-b_4 \ r_{26}) + \exp(-b_4 \ r_{35}) + \exp(-b_4 \ r_{45}) \ ] \tag{11}
$$

The set of constants obtained by Matsuoka et al. is, in A and kcal/mole:

a_1 = 1.088.213 a_2 = 666.3373 a_3 = 1455.427 a_4 = 273.5954 b_1 = 5.152712
b_2 = 2.760.844 b_3 = 2.961895 b_4 = 2.233264 q^2 = 170.9389.

While stressing that any constant appearing in the analytical potential is only a fitting parameter, we may note that the dipole moment calculated using the above parameters is 2.19 D, to be compared with the dipole moment for a single water molecule, 1.85 D.

With the above potential, a Monte Carlo simulation on the liquid water structure was recently performed[32]. In this study, 343 water molecules are considered in a cube with periodic boundary conditions, the assumed water density is equal to the experimental density at 25°C (0.03334 molecules/A^3). The size of the cube is thus $(21.73)^3$ A^3 and is believed to be big enough that the long range water-water interactions are sufficiently accounted for. In the simulation the total number of configurations generated (after rejecting the initial 500.000 configurations) is 600.000; about half of them are rejected in the Markov chain[33].

The average distribution of molecules in a liquid is usually represented by radial distribution functions. The computed oxygen-oxygen radial distribution functions, $g_{OO}(r)$, $\Gamma_{OO}(r) = 4\pi\rho r^2 \ g_{OO}(r)$, and $N_{OO}(r) = \int_0^r \Gamma_{OO}(R)dR$ are shown in Fig. 16, together with the experimental results of Narten et al.[34]. Here is to be noted that, in deriving three radial distributions, g_{OO} g_{OH} and g_{HH}, from experimental x-ray and neutron scattering data, a model for the average orientation of pairs of near neighbouring molecules had to be assumed. (Narten had at hand only two methods involving sufficiently different values for the atomic scattering amplitudes).

It is clear from Fig. 16 that the agreement between the simulated and experimental g_{OO} is very satisfactory. This is clearly shown by analyzing the height and position of the first peak : the first peak in the simulations occurs at an oxygen-oxygen distance 2.83 A with height value of 2.48, to be compared with the x-ray data of 2.82 A and 2.31 for distance and height, respectively. Lie and Clementi[35] obtained in their Monte Carlo simulation, using only radial correlated pair potentials, the peak 2.78 at 2.78 A, whereas Stillinger and Rahman obtained, using an improved empirical potential in a molecular dynamics simulation, 3.09 at 2.85 A[36]. The two broad maxima observed experimentally at 4.5 A and 6.5 A are also well reproduced in the simulation. The agreement between the simulated and the experimental result is even more evident by taking into account the volume element, as can be seen from Figs. 16(b) and 16(c). It is interesting to note that all simulations of liquid water carried out up to this time, including the present work, fail to reproduce exactly the distinct, albeit small, peak observed experimentally at 3.75 A, an indication that the peak is probably a cut off ripple caused by the finite termination of the upper convolution limit.

In the radial distribution functions of Narten and Levy[34], the maximum of the first peak shifts gradually from 2.82 to 2.24 A as the temperature is increased from 4 to 200°C. A similar trend is expected to be found in our simulation since, as noted previously, the minimum energy configuration for the pair potential used in the present work is 2.87 A, whereas the nearest neighboring oxygen-oxygen distance found in liquid at 25°C is 2.83 A.

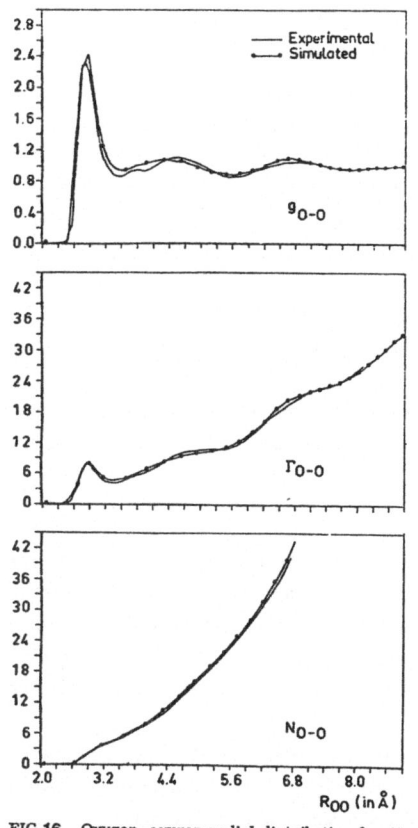

FIG.16 Oxygen–oxygen radial distribution functions.

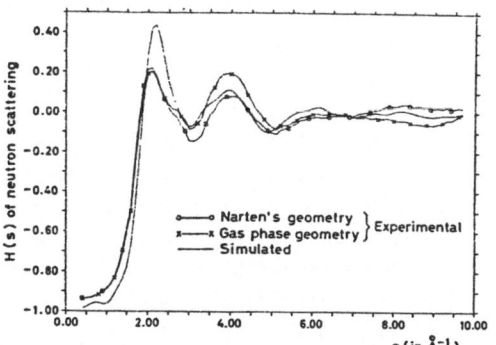

FIG.17 Structure function $H(s)$ from neutron scattering (see text for the explanation of the two different geometries used for the water molecule in constructing the experimental curves).

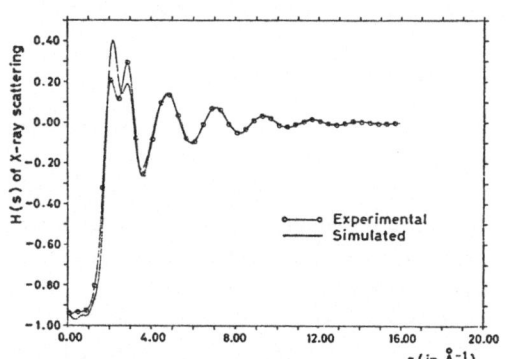

FIG.18 Structure function $H(s)$ from x-ray scattering.

In Table 2 we numerically compare the radial function obtained with the results of the most recent, accurate molecular dynamic simulation of Stillinger and Rahman. It should be noted that any difference rests mainly in the differences in the potential functions used, but not the simulation techniques. The potential used by Stillinger and Raham, called ST2, is of empirical nature and predicts also a nearly linear hydrogen-bonded structure with bonding energy -6.8 kcal/mole as the most stable dimer configuration. All results of the ST2 potential reported in Table 2 are interpolated from the original paper between 4 and 41 °C.

Table 2 - Comparison of radial distribution functions at 25° C

		1st max		1st min		2nd max		2nd min	
g_{OO}	ST2	2.85	3.09	3.53	0.72	4.70	1.13	5.8	0.80
	CI	2.83	2.46	3.53	0.94	4.25	1.08	5.6	0.89
	Exptl	2.83	2.31	3.45	0.85	4.53	1.12	5.6	0.86
g_{OH}	ST2	1.90	1.38	2.50	0.31	3.40	1.60	4.60	0.92
	CI	1.90	1.08	2.55	0.28	3.35	1.68	4.85	0.90
	Exptl	1.90	0.80	2.45	0.50	3.35	1.70	−f	−f
g_{HH}	ST2	2.50	1.50	3.10	0.78	4.00	1.15	5.40	0.96
	CI	2.50	1.40	3.10	0.86	3.90	1.20	5.50	0.93
	Exptl	2.35	1.04	3.00	0.47	4.00	1.08	−f	−f

Since the radial distribution functions are indirect experimental data, it is desirable to calculate the x-ray and neutron scattering intensities from our simulated radial distribution functions and thus to be in a position to compare directly with the corresponding original experimental data.

The coherent scattering intensity $I(s)$ for x-ray, neutron, or electron scattering can be treated in the same way and, neglecting the surface scattering, is given, in number of electrons per molecule, by

$$I(s) = (F^2) + \left(\sum_\alpha \sum_\beta f_\alpha f_\beta \right) \int_0^\infty 4\pi r^2 \rho [g_{\alpha\beta}(r) - 1] \frac{\sin(sr)}{sr} dr,$$

where

$$(F^2) = \sum_i \sum_j f_i f_j \exp(-b_{ij} s^2) \frac{\sin(sr_{ij})}{sr_{ij}}$$

is the scattering intensity from one independent molecule averaged over all orientations, depending on the average interatomic distances r_{ij} and their mean-square variations $2b_{ij}$. Note that indices i and j are used to refer to atoms belonging to the same molecule, while α and β to different molecules. f_α (or f_i) is the static coherent scattering amplitude for atom α (or i). ρ is the experimental (bulk) density of water and g (r) the radial distribution function for the pair α and β. s is related to the scattering angle 2θ and the wavelength λ of the incident beam by the following equation:

$$s = 4\pi \sin\theta / \lambda .$$

The scattering amplitude f_a is independent of s in the case of neutron scattering and is a function of s for x-ray scattering.

In studying the structure of liquid, however, it seems more meanful to subtract out the intensity due to the scattering of a single independent molecule; hence, we can introduce a structure function $H(s)$ defined as

$$H(s) = \frac{I(s) - (F^2)}{(F^2)}$$

The denominator is used for normalization, so that $H(s)$ approaches 0 as s-0 if there is no structure (i.e., idea gas) and -1 if the structure is highly ordered (e.g., perfect rigid crystal). We note that the definition used here is different from that of Narten et al.[34], and is believed to be more instructive in presenting a structure function for nonsimple liquids, since all interference patterns in $H(s)$ are then solely caused by the deviation of g from unity. At large angles only independent molecular scattering is observed, and hence $H(s)$ approaches zero at large s.

The neutron atomic scattering amplitudes used in this work are $f_0 = 0.566 \times 10^{-12}$ cm[37] and $f = 0.67 \times 10^{-12}$ cm[38]. We assume that the structure of liquid H_2O is the same as that of liquid D_2O in order to compare the simulated neutron intensity with experiment. The atomic x-ray form factors, a function of s, are extracted and interpolated from the tabulation of Narten and Levy[34]. Fig. 17 compares the simulated x-ray structure function with the experimental results. The agreement seems to us to be very satisfactory. The unique double peaks observed experimentally in the range $s \cong 2$ A^{-1} and $s \cong 3$ A^{-1} have been found also in the simulation, although the computed left peak is too high, whereas the right one is too low. Comparing with the simulated results of Lie and Clementi[35], where no angular correlations were included in the potential, we can notice not only qualitative but also quantitative improvements in the present work. The structure functions constructed from the neutron scattering intensities of the simulation and the experiment[34] are compared in Fig. 18. The agreement is less satisfactory but still moderately good. One reason for the discrepancy may stem from the fact that the neutron scattering lengths for the oxygen and deuterium atoms are about the same, hence a precise knowledge of the dimensions of the water molecule is necessary in untangling the intramolecular scattering from the experimental total intensity. This difficulty is practically not present in the x-ray scattering of liquid water, since the scattering is dominated by the oxygen atom. Two different geometries for the water molecule are used to subtract the single molecule scattering from the experimental results of Narten in Fig. 16: the first one is that obtained from the gas phase by Benedict et al.[39] with $b_{OH} = 0.0022$ A^2 and $b_{HH} = 0.0066$ A^2 from Shibata and Bartell[40], the second that obtained by Narten[34] by least squares fitting to the observed data and with the assumption of a model. It should therefore be noted that the OD distance found by Narten is ˜0.02 A shorter than the equilibrium distance in D_2O vapor, whereas the DD distance is ˜0.02 A longer. However, Narten suggested that differences should not be accepted as real unless confirmed by a more refined neutron experiment. Another source of discrepancy between the simulation and experiment may lie in the following different definitions of the "structure" as "observed" by the simulation and the x-ray and neutron experiments. As is well known, the molecules in a liquid undergo rotational and translational displacements and more frequent oscillations. The structure observed in the coherent x-ray or neutron scattering is called, according to Eisenberg and Kauzmann[41], diffuse-averaged or D structure, which can be regarded as either a time average or a space average of different vibration-averaged structures. In our simulation, however, no vibrational average of each displacement generated by a Markov chain was taken into account.

Despite all these problems in interpretation and comparison it seems clear from Fig. 16 that the intra-molecular scattering affects almost exclusively the region for $s > 2.5$ A$^{\prime 1}$ and that the simulated positions of the scattering maxima and minima agree well with experiment, but the simulated intensities less so. Since the zero angle scattering is connected with the isothermal compressibility, we postpone its discussion to the next section.

We are of the opinion that starting from first principles, one can readily simulate equilibrium properties of a quite complex system, like liquid water, to a high degree of accuracy. Although three-body (or even four-body) interactions (and possibly quantum effects) have frequently been suggested to be important in the study of the structure of water, it seems, judging from the results presented here, that they contribute only nominally to the pair distribution functions and heat capacity. We are presently working towards a quantitative definition of the importance of three-body effects.

2.6 References

1   H. Popkie, H. Kistenmacher and E. Clementi, J. Chem. Phys. 59, 1325 (1973).
2   E. Clementi and H. Popkie, J. Chem. Phys. 57, 1077 (1972).
3   See, e.g., K. Morojuma and L. Pedersen, J. Chem. Phys. 48, 3275 (1968) and J. Del Bene and J.A. Pople, J. Chem. Phys. 52, 4858 (1970).
4   J.D. Bernal and F.D. Fowler, J. Chem. Phys. 1, 515 (1933).
5   J.S. Rowlinson, Trans. Faraday Soc. 47, 120 (1951).
6   A. Ben Naim and F. Stillinger, "Aspects of the Statistical Mechanical Theory of Water", in Structure and Transport Processes in Water and Aqueous Solutions, edited by R.A. Horne (Wiley, Interscience, New York, 1972).
7   G.F.H. Diercksen, Theoret. Chim. Acta 21, 335 (1971).
8   D. Hankins, J.W. Moscowitz and F. Stillinger, J. Chem. Phys. 53, 4544 (1970). See also: D. Neumann and J.W. Moscowitz, J. Chem. Phys. 49, 2056 (1968).
9   E. Clementi. H. Kistenmacher and H. Popkie, J. Chem. Phys. 58, 2460 (1973).
10 See, for example, P.A. Flory, "Statistical Mechanics of Chain Molecules" Interscience Publishers, John Wiley, New York (1969).
11 D. Hankins, J.W. Moscowitz and F.H. Stillinger, Chem. Phys. Letters 4, 527 (1970) and J. Chem. Phys. 53, 4544 (1970).
12 B.R. Lentz and H.A. Scheraga, J. Chem. Phys. 58, 5296 (1973).
13 H. Kistenmacher, G. Lie, H. Popkie and E. Clementi, J. Chem. Phys.
14 D.E. Schreiber, IBM Rsearch, RJ 499 (1968).
15 C.K. Johnson, ONRL Report 3794 (program ORTEP).
16 M. Van Thiel. E.D. Becker and G. Pimentel, J. Chem. Phys. 27, 486 (1957).
17 P.A. Kollman and L.C. Allen, J. Chem. Phys. 51, 3286 (1969).
18 J.P. Chandler, Q.C.P.E., document 66, University of Indiana, USA.
19 J. Del Bene and J.A.Pople, J. Chem. Phys. 52, 4858 (1970) and 58, 3605 (1973).
20 A. Ben Naim, J. Chem. Phys. 57, 3605 (1972).
21 H. Kistenmacher, H. Popkie, E. Clementi and R.O. Watts, J. Chem. Phys. 60, 4455 (1974).
22 E.P. Wigner, Phys. Rev. 46, 1002 (1934).
23 J.A. Barker and R.O. Watts, Chem. Phys. Letters 3, 144 (1969).
24 D. Eisenberg and W. Ka uzmann "The Structure and Properties of Water", Oxford University Press, p.44, Table 2-3 (1969).
25 C.W. Kern and M. Karplus, Chapter 2 in "Water, a Comprehensive Treatise", Vol. I, F. Frank, Plenum Press, New York (1972).
26 F.H. Stillinger, J. Chem. Phys. 57, 1281 (1972).
27 A.H. Narten, J. Chem. Phys. 56, 5681 (1972).
28 G.H.F. Diercksen, N.P. Kraemer and B.O. Ross, Theor. Chim. Acta, 36, 249 (1975).
29 W. Kolos (private communication).
30 O. Matsuoka, E. Clementi and M. Yoshimine, J. Chem. Phys.
31 C. Edmiston and K. Ruedenberg, Rev. Mod. Phys. 34, 457 (1963).
32 W.S. Benedict, N. Gailar, and E.K. Plyler, J. Chem. Phys. 24, 1139 (1956).
33 J.A. Barker and R.O. Watts, Chem. Phys. Letters 3, 144 (1969).
34 A.H. Narten and H.A. Levy, J. Chem. Phys. 55, 2263 (1971).
35 G.C Lie and E. Clementi, J. Chem. Phys. 62, 2195 (1975).
36 Interpolated between 10°C and 41°C.
37 G.E. Bacon, Acta Crystallogr. A 25, 391 (1969).
38 C.G. Shull. M.I.T. compilation, February (1971).
39 W.S. Benedict, N. Gailar, and E.K. Plyler, J. Chem. Phys. 24, 1139 (1956).
40 S. Shibata and L.S. Bartell, J.Chem. Phys. 42, 1147 (1965).
41 D. Eisenberg and W. Kauzmann, "The Structure and Properties of Water", Oxford U. P., New York (1969).

PART 3 - Coordination Numbers and Solvation Shells

3.1 Introduction

One should distinguish between coordination numbers and hydration numbers. Following Bockris[1] the coordination number of an ion is the number of molecules of water in the immediate neighbourhood of the ion, and depends on the distance between the molecule of water and the ion. The hydration number is based on the dynamical behaviour of the water molecules in solution, that at a given temperature move with the ion as a unit. The Monte Carlo method and the scattering methods therefore deal with coordination numbers, other methods like measurements of viscosity, conductivity and compressibility deal with hydration numbers.

In this section we shall consider "coordination numbers" and the solvation shell radii; it is clear from the above definition that the coordination number is meaningful only if defined relative to a radius (or in general an exactly defined volume) of a sphere centered at the ion. The definition of the volume shape is obvious for a single ion in solution, but less so for an ion-pair at short interionic distances, as we shall indicate in the second section of this Part 3. A problem not sufficiently studied is the definition of the volume relative to the coordination number for a more complex chemical, for example for a -COOH, -NH$_2$ etc. group. This problem will be discussed in the last section of this Part 3, where we shall consider all the naturally occurring aminoacids. In the field of biomolecules in solution there is a vast amount of work and we wish to point out to the recent review paper of A. Pullman and B. Pullman[2]. We note that the opportunity to obtain transferable potential functions expressed as an analytical potential seems to have not been considered by several authors and this prevents the possibility to introduce temperature effects and statistical distributions; however, the above referred review[2] represents a very nice summary of a contribution to the problem of the orientation of one molecule of water relative to biological molecules. We shall indicate that by introducing the water-water interaction (see later sections 3.2 and 3.3), one can obtain a statistically correct description of ions in solution at a given temperature. Biomolecules interacting with water is the subject of the last section (3.4) of this review; since the interaction of biomolecules with water is expressed via analytical potentials, as previously done for the water-ions interactions, it follows that not only we can obtain a statistically correct description of biomolecule in solution, but also that we can make use of the same computer programm tested for liquid water and ion-water.

3.2 Structure of Water around Ion-Pairs

In this section we present a study of ion pairs in a cluster of two hundred water molecules using Monte Carlo calculation[3]. To enable us to study the detailed distribution of the water molecules around the ion pair, the ions were held at three fixed distances apart, namely 6 A, 8 A and 10 A, and the calculations carried out at 298 K; one further calculation[4] was performed for the lithium fluoride system at a temperature of 500 K. Although there was ample opportunity for the water molecules to evaporate, a distinct cluster was formed and remained stable[4].

There are three interactions of importance in our studies, namely the water-water interaction, the cation-water interaction and the anion-water interaction. As we are fixing the relative positions of the ions, the ion-ion interaction potential does not enter the study at this time. Accurate Hartree-Fock interaction potentials exist[5-6] for all the required interactions and furthermore recent Hartree-Fock calculations have considered the $H_2O - Li^+ - F^-$ three body interaction[7]. Thus these calculations are in effect ab initio studies of the molecular complexes and do not rely upon any parameters adjusted empirically to reproduce experimental properties of the ionic clusters.

A Monte Carlo program originally developed by Barker and Watts for liquid water[8] was modified to simulate a finite system of water molecules around an ion pair held at fixed distance apart. The cluster was contained in a cubic box with the anion at position $(-x,0,0)$ and the cation at $(x,0,0)$, whose size was such that if the water molecules filled the box uniformly the density would be one-eighth of the density of liquid water at 298 K and at atmospheric pressure. As we shall see the various interactions in the system were sufficiently strong that the cluster did not disperse. An initial configuration was chosen, for all three ion-ion separations, in which the water molecules were placed on an ice l-c lattice surrounding the ion pair. The effect of this starting configuration was removed by using an annealing procedure reported for the pure water calculations[8]. Once the system was annealed further 5×10^5 configurations were generated to ensure that the system had reached equilibrium. These initial configurations were discarded and all results reported here were obtained by averaging over at least 10^6 configurations.

To ensure that the system did in fact remain as a cluster and did not evaporate to fill the box uniformly, the average numbers of oxygen atoms at various positions along the x, y and z axes were calculated. By comparing Monte Carlo simulations performed at different temperatures, it appears that there is a smooth transition from a tightly bound complex to a looser association and it is possible that the water-alkali halide cluster does not show a clear liquid-vapor phase transition[4]. However, it may be necessary to make a more detailed study before confirming this suggestion.

In the theory of dense fluids it is usual to report the structure in terms of the radial distribution function of the system. To examine the detailed structure of the ionic solution, we have generalized this function to obtain the density, or single particle distribution function for the oxygen and hydrogen atoms given the fixed ion-pair positions. The result is a distribution function

$$f_i(x, r) = \frac{\langle N_i \{(x, x+ \delta x), (r, r+ \delta r)\} \rangle}{2 \pi^r \delta r \delta x} \tag{1}$$

where the quantity averaged is the number of atoms of type i (oxygen or hydrogen) having their coordinate in the range $(x, x+\delta x)$ and being within the distances $(r, r+\delta r)$ of the x-axis, where $r^2 + y^2 = z^2$. This quantity was calculated over a matrix of (x, r) values for both the oxygen and hydrogen atoms for all four systems. In Fig. 1 the overall structure of the clusters is represented by reporting the statistical average of the population of the atoms at various positions along the x axis for the case of an inter-ionic distance of 10.0 A.

The vertical scale is such that $10 N_1$ is the number of molecules per unit length. For all cases, the cluster center appears to be un-symmetrically displaced along the x-axis and no striking regularity can be detected either by comparing different ion-pairs at the same anion-cation separation or by comparing, for a given ion-pair, different anion-cation separations. We find that the length of the cluster along the x-axis increases with increasing distance of the two ions, but this feature is clearly expected. In addition we note that for the case of NaCl the cluster's left and right boundaries are more symmetrically located than in any other .case. In the case of NaF a cleavage can be detected which is more strongly marked than in the case of the other two fluorides at any distance. The cluster character is retained at all interionic distances at the chosen temperature, namely there is a very well defined beginning and a very well defined end at the two extremities of the cluster. The fine structure of the cluster is another general characteristic that emerges from the probability distribution reported in Fig. 1. This fine structure, having an interval, from maximum to maximum, of about 3 to 4 A, indicates that the whole system of 200 molecules of water and the two ions is highly structured even at room temperature. Three-dimensional pictures obtained by video display of a number of specific configurations, rather than by considering the statistical distribution of the water molecules, reveal "filaments" of water molecules, where one molecule is hydrogen bonded to the other, departing from the ions and extending towards the cluster surface. The fine structure of the cluster is due to these filaments, or polymers, of the water.

Since this feature has been noted over and over, either using different potentials or different temperatures or different ions, we are of the opinion, it represents a real feature of the water organization around ions, and with a reasonable extrapolation, around ionic groups (for example, a carboxylic group). Considering the water arrangement about ions, traditionally one talks about hydration shells. Equally reasonably one could talk of water polymers departing radially from the ion and having a characteristic length and a stability that decreases with increasing distance from the point charge at the ion. After a given number of units (water monomer) the order introduced by the ionic field, (reflected in the water polymer formation) is counterbalanced by the thermal motion (disorder) of the liquid. The concept of solvation shell in part originates from considering the liquid water as a continuum medium, and therefore neglecting to some extent the molecular aspect of the water molecules in the liquid. The concept of water polymers can help a complementary description, in re-stating the molecular, the geometrical and the discrete aspect of the liquid.

The radial distribution functions of interest in the cluster studies are those for the positive or negative ion (i) and for the hydrogen or oxygen atoms (A), namely

$$f_i^A \ (x, r) \ = \ \frac{\langle N_i \{(x , x + \delta x) , \ (r, r + \delta r)\} \rangle}{2 \ \pi r \ \delta \ x \ \delta \ r} \tag{2}$$

where the oxygen (or hydrogen) is in the range $(x, x+\delta x)$ and within $(r, r+\delta r)$ of the axis; $(r^2 = y^2 + z^2)$.

Fig. 2 reports the probability plot for the oxygen distribution, $f_i^O \ (x, r)$, Fig. 3 the plot for the hydrogen distribution, $f_i^O \ (x, r)$. There is not only evidence of a first hydration shell, but also of a second hydration shell.

The simplest and somewhat naive definition of the coordination number for a given ion in aqueous solution is simply the number of water molecules that surrounded that ion within a defined volume, for example a sphere, having the ion at its center. A more physical definition should explicate the relationship of the coordination number with temperature and with the presence or absence of other ions. In this section we have presented a number of computer experiments (all done at the same temperature) where we have obtained the statistical distribution of water molecules around a pair of ions, at various inter-ionic separations.

We can integrate over the space coordinates for the correlation functions $f_i^O \ (x, r)$ and $f_i^H \ (x, r)$, previously defined; the integration is for a volume having the ion (indicated by the subscript i) as its center. The function obtained by integrating $f_i^O \ (x, r)$ from r=0 to a finite value is designated as N_O, and the function obtained by integrating $f_i^H \ (x, r)$ from r=0 to a given value is designated as N_H. Clearly, N_H and N_O can be used for a definition and for a determination of the coordination numbers, since the function N_H or N_O corresponds to the number of hydrogen atoms or to the number of oxygen atoms contained in a defined spherical volume having the ion as origin. In Fig. 4 we report N_H and N_O for the negative ions F⁻ and Cl⁷ in presence of a given counter-ion computed for the different ion-pair separations previously discussed. Since for the negative ion, one of the hydrogens of a water molecule is hydrogen bonded to the anion, we report two curves for N_H, one corresponding to the hydrogen nearer to the anion (hydrogen bonded), the second for the hydrogen further away and not hydrogen bonded to the anion. In Fig. 5 we report the values of N_H and N_O for the Li⁺, Na⁺ and K⁺ cations in presence of F⁻ or Cl⁻; in this case, since the hydrogens are not hydrogen bonded to the cations, we report only one function for the hydrogen distribution. It should be pointed out, however, that a difference in the probability distribution for the two hydrogens of a water molecule can be detected in some of the reported graphs describing N_ρ for hydrogen atoms around positive ions; in particular for the case of the Li⁺ cation a splitting of the maximum probability into two probability peaks can be noticed. On the other hand, we feel that it is likely simply an effect due to the presence of the anion in the ionic pair studied, since the effect is more pronounced the nearer the inter-ionic separation.

Since a plateau for a given range of r in the N_H and N_O curves indicates that the integral value remains constant for that range of r, the detection of such plateau is an indication that a "hydration shell" is completed and thus provides us with a quantitative value of the hydration shell radius. Thus, in principle, from the graphs of N_H and N_O we could obtain the coordination numbers and the shell radius. From inspection of the Figs. 4 and 5, it appears, however, that it might not be sufficiently unambiguous how to determine the radius. Therefore, we report in Figs. 6 and 7 the derivatives of N_H and N_O; indicated as N_ρ: the vertical scale is such that $N_\rho/4\pi r^2 (\Delta r)$ is is molecules per unit volume (A^3). From Figs. 6 and 7 we can obtain the radii of the first and second hydration shells in a simple way. First, we must realize that a shell has an intrinsic thickness, therefore it is defined by an initial radius (inner edge) and a final radius (outer edge). The initial radius corresponds to a maximum in Figs. 6 and 7 of a given distribution, the final radius corresponds to the minimum of the same distribution; therefore, we can define an average radius as the arithmetic mean of the initial and of the final radius. Once the radius is determined the computation of the coordination numbers is immediate, since it can be read directly from the data in Figs. 4 and 5. Second, one must distinguish between the anions and the cations; for the anions there is a distinct distribution for the inner hydrogen (hydrogen bonded) and for the outer hydrogen of H_2O; for the cation the two hydrogen distributions are over imposed. Thus in this work, to properly interpret our simulated data, we must consider a shell as an organized structure having a finite thickness, where the shell is properly defined; between shells there is a region of statistically rather disordered structure. The coordination numbers are as usual the number of molecules of water in the volume defined by the shell radius. For example, in the case of Li^+ (see Table 1) the organized structure of the first shell is within two radii of 2,2 A and 3.3 A (giving an average of 2.7 A). The second shell is statistically well defined from a radius of 4.7 A to a radius of 6.2 A; in the region 3.3 A to 4.7 A the density distribution is low and statistically poorly defined (note that the thickness of a shell, defined as the interval between the two radii is of the order of about 1 A (for cations, and larger for the anions as expected on hydrogen bonds considerations).

Equivalently we can define the radius of the second hydration shell. However, in this case some additional care must be used, due to the fact that we have considered finite inter-ionic distances. For a given radius of the volume surrounding an ion (e.g. the anion) the water included in the sphere not only "belongs" to that ion but might belong, in addition, to the second ion (e.g. the cation). This is a difficulty we encounter in defining unambiguously the coordination number for ions in clusters containing two ionic species (clearly, for highly diluted solutions this problem does not arise, since the solvation shells do not overlap).

In order to avoid this ambiguity, we define the coordination numbers by limiting the integration of the function f(x, r) to the volume either to the left (for the cation) or to the right (for the anion) with a perpendicular to the x-axis (the inter-ionic axis) and bisecting it at $x = 0.0 A$.. In this way the ambiguity is removed to a large extent.

In Table 1 we summarize the values obtained with the above procedure, for the radii and coordination numbers of the first and second shells, reporting only the average values. From the data in Table 1, we notice that the errors in the determination of the radius are relatively small (up to 10%), with exception of the K^+ cation, where we have not been able to determine unambiguously the radius of the second hydration shell. The error in the determination of the coordination number is between 10% and 25%, i.e., much less precisely determined than the radii. It was one of our aims for this work to study the influence on the coordination numbers and radii of the second ion in the ion-pair situation; the error limit of our computation is too large to see such effect. Thus, we can preliminarily conclude that the perturbation of the second ion is likely smaller than the above given errors, for ion-ion distances larger or equal to 8 A, but more work is needed to settle the problem.

Table 1 — Coordination numbers and hydration shell's radii (A) for selected ions

Ion	First shell radius	First shell coordin.	Second shell radius	Second shell coordin.
Li^+	2.7 ± 0.1	5.4 ± 0.7	5.1 ± 0.4	13.9 ± 2.7
Na^+	3.4 ± 0.3	6.0 ± 1.1	5.9 ± 0.3	17.1 ± 2.8
K^+	4.0 ± 0.3	7.2 ± 1.2	(5.4 ± 0.3)	(12.6 ± 4.0
F^-	3.0 ± 0.5	4.5 ± 0.7	4.7 ± 0.4	15.4 ± 1.7
Cl^-	3.9 ± 0.5	5.1 ± 0.8	6.2 ± 0.5	17.9 ± 2.4

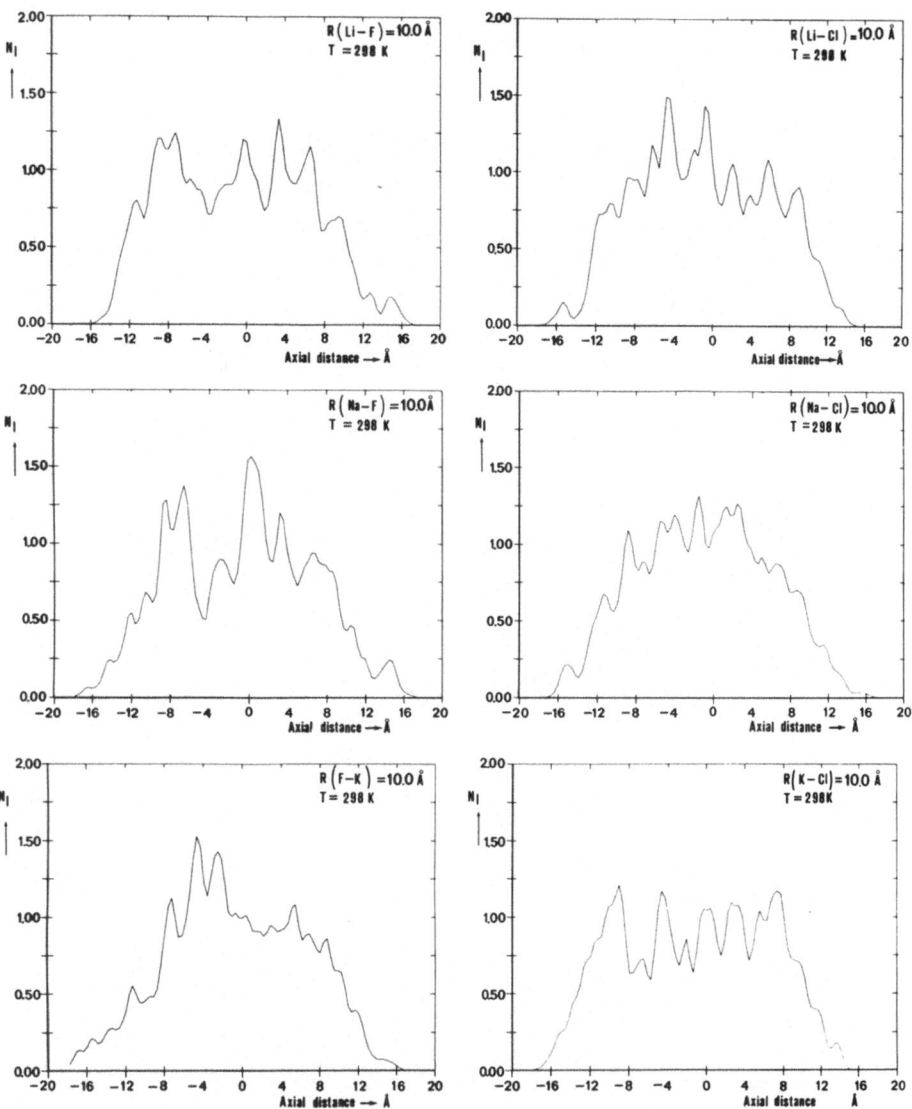

Fig. 1 - Cluster envelope projected on the X-axis for the whole set of ion pairs. The position of the ion pair is marked on the X-axis. The distance between the ions is 10.0 Å.

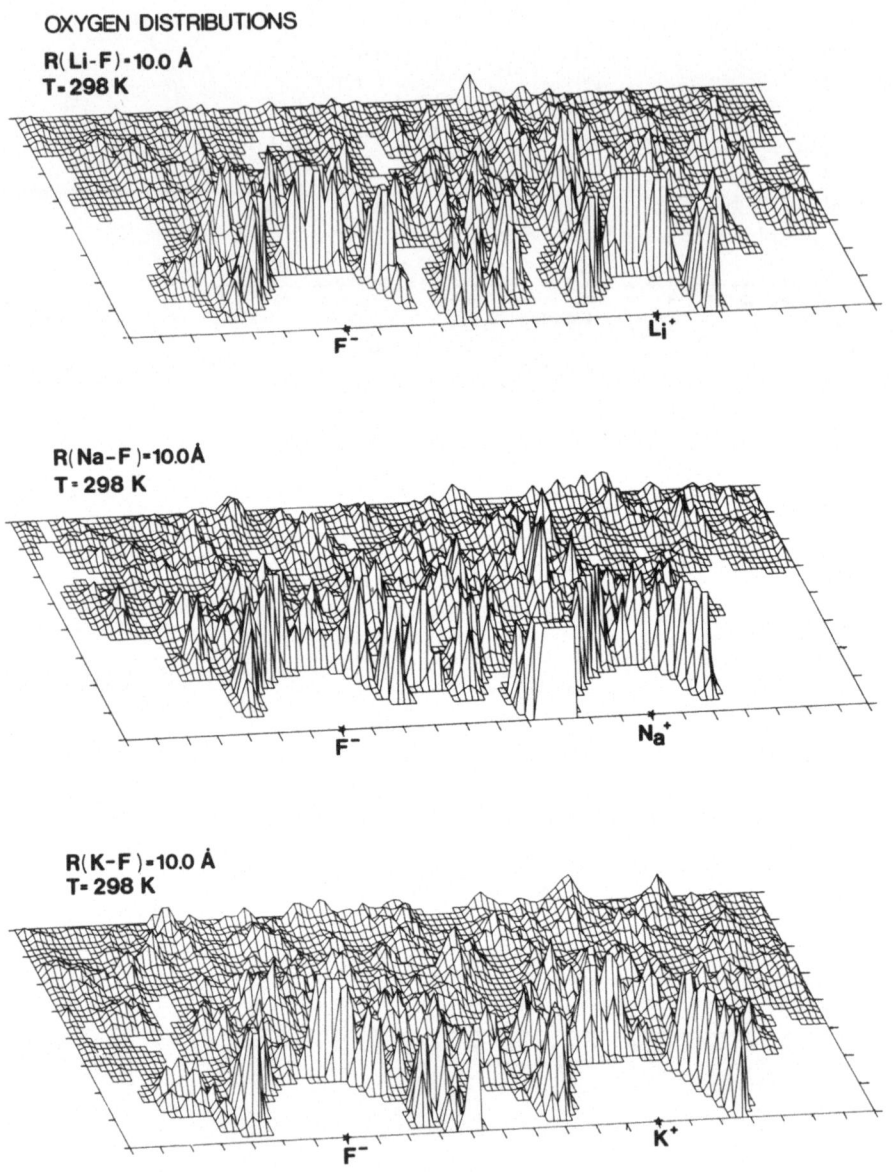

Fig. 2 - Oxygen distribution: three dimensional representation for the most relevant structure of the cluster, relative to the ion pairs LiF, NaF and KF at an interionic separation of 10 Å.

HYDROGEN DISTRIBUTIONS

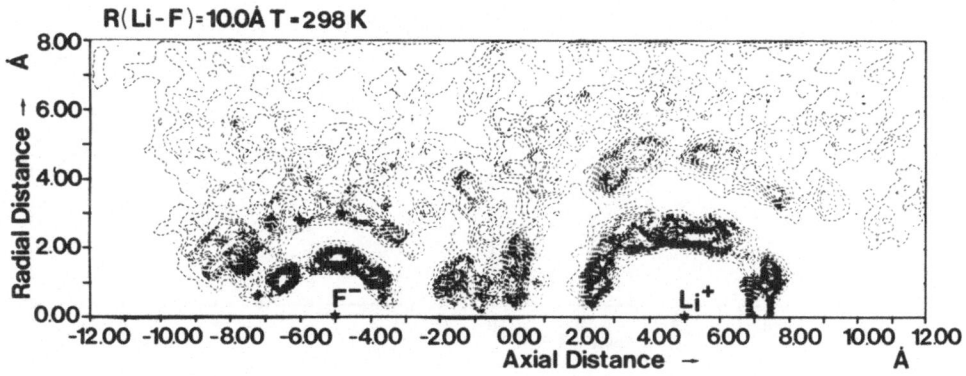

R(Li-F)=10.0Å T-298K

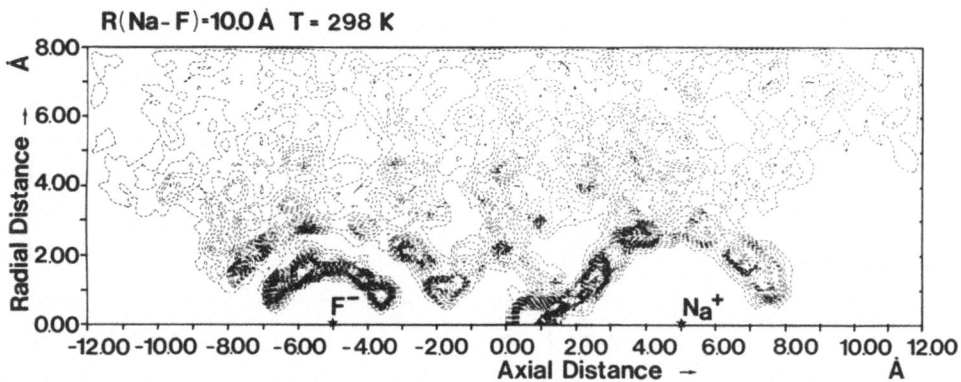

R(Na-F)-10.0Å T-298K

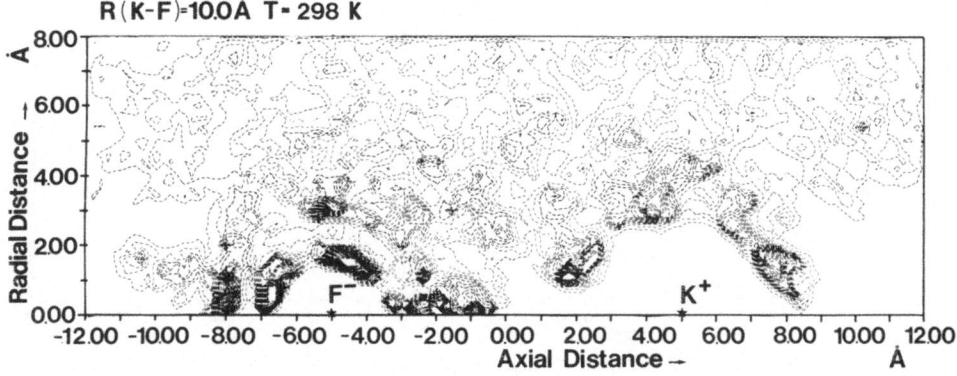

R(K-F)=10.0Å T-298K

Fig. 3 – Hydrogen distribution contour maps for the most relevant structure of the cluster. The maps are relative to the ion pairs LiF, NaF and KF at an interionic separation of 10.0 Å.

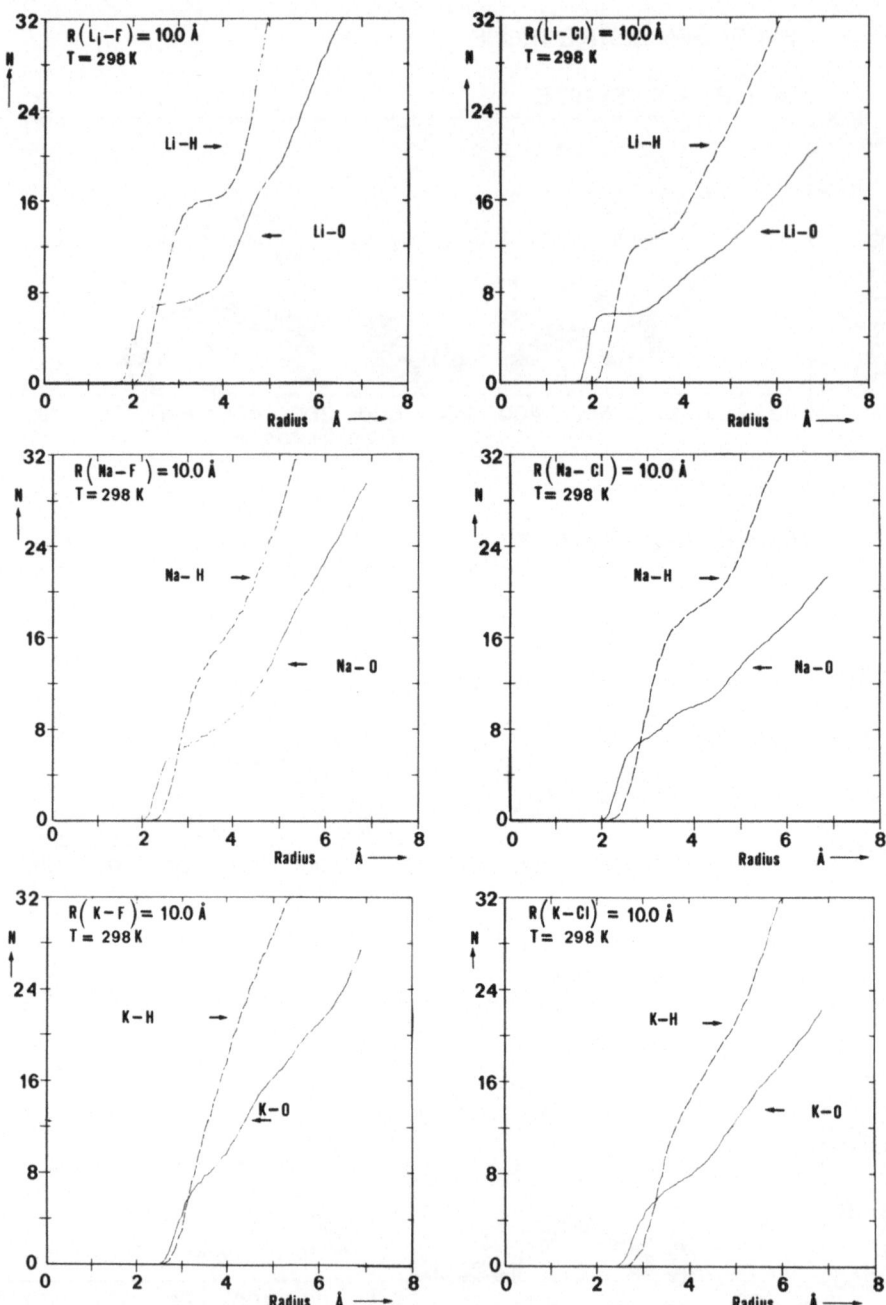

Fig. 4 – Integral of local density for the sum of the two hydrogens of water and for oxygen atoms relative to the cation placed at the origin. The six pictures are relative to the whole set of ion pairs at an interionic separation of 10.0 Å.

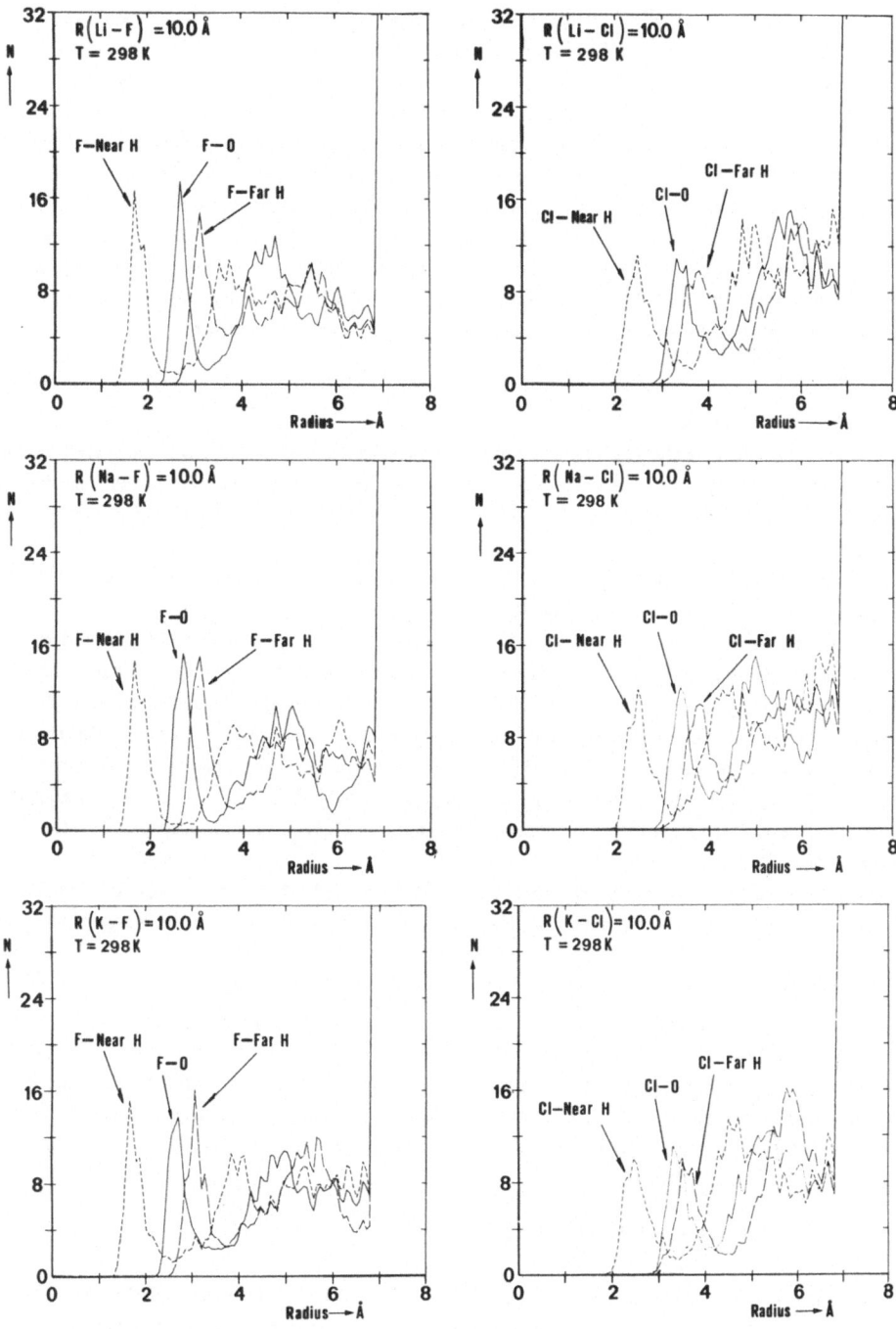

Fig. 5 – Integrand value of the local density distributions given in Fig. 4.

A comparison with experimental determinations of the coordination number reveals that our determinations are quite useful, since all experimental determinations depend heavily on the model that is assumed for the solvent around the ion, and none are direct determinations of the number of molecules around a small volume element of solvent.

For the Li^+, Na^+ and K^+ ions the coordination numbers experimentally reported are 4.5, 4.5 and 3.8, respectively (compressibility data[1]), 5, 4 and 3, respectively (entropy data[9]), 2.5, 4.8 and 1.0, respectively (density data[10]), 5.5, 7.0 and 6.0, respectively (compressibility data[1]), 5±2, 3±1.2, and 1±0.4, respectively (n.m.r. data[12]), 3.4, 4.6, and 4.6 (n.m.r. data[13]). This extended spread of "experimental data" for the cations is similar to what reported for the anions: for F^- and Cl^- the reported experimental coordination numbers are 4.0 and 2.2 respectively (compressibility[1]), 5 and 3 (entropy[9]), 2.0 and 1.0 respectively (compressibility[11]), 9.9±2 (n.m.r. data[14]). This list of experimental numbers is not complete but, in our opinion, is indicative of the situation from the experimental side.

3.3 The Three-Body Effects

In this section we consider the H_2O-Li-F complex. The choice of this complex was motivated by a number of considerations. Firstly, since we wish to use the potential function derived here as the starting point in a study of ionic solutions, it is essential to consider a cluster having a water molecule in interaction with both a positive and a negative ion.

Secondly, we wish to study the importance of the three-body correction term, i.e., to study the reliability of the pairwise additivity assumption in the Hartree-Fock framework.

Because this section presents a detailed study for a system of "three bodies" (Li^+, F^-, and H_2O), some discussion of the energy analysis for such a system is in order. Two analyses can be made. The first is the classically-motivated, pairwise additivity analysis which relates the total three-body interaction to a sum of two-body interaction terms, with a residual "three-body effect". The second procedure uses the quantum-mechanical bond energy analysis scheme[15] where the total Hartree-Fock energy is partitioned into one-body, two-body, and three-body terms. Both analyses will be discussed and compared and, in particular, we shall establish a direct relationship between the classical decomposition and the quantum-mechanical BEA.

To fix our ideas, suppose we denote the total energy of the three-body system as $V(1, 2, 3)$ and the energy of the ith isolated one-body system as $V(i)$; then, we define the total stabilization energy in the usual way as

$$E(s, tot) = V(1, 2, 3) - \sum_i V(i) \tag{3}$$

Note that this energy is negative for a three-body complex which is more stable (lower in energy) than its isolated constituents.

A common assumption in many-body theory is the pairwise additivity assumption; here, the interaction energy of a three-body system is represented as the sum of two-body interactions. In terms of the stabilization energy, the assumption of pairwise additivity can be expressed formally as

$$E(s, 2) = \sum_{i < j} V(2, ij) \tag{4}$$

Here, $E(s, 2)$ is a "two-body stabilization energy",

$$V(2, ij) = V(ij) - V(i) - V(j)$$

is the stabilization energy of the isolated pair ij, and $V(ij)$ is the total energy of the isolated pair ij. In general, $E(s,2)$ need not be the same as the total stabilization energy of the three-body system; accordingly, we may define a three-body stabilization energy

$$E(s, 3) = E(s, tot) - E(s, 2) \tag{5}$$

The energy difference $E(s,3)$ defines the classical nonadditive contribution to the total stabilization energy of the system.

In the BEA, the total energy is decomposed into one-body, two-body, and three-body terms as follows

$$V(1,2,3) = \sum_i \&(i) + \sum_{i<j} \&(ij) + \sum \&(ijk) \tag{6}$$

with the $\&(i)$, $\&(ij)$, and $\&(ijk)$ defined as in ref.[14] This alternate, formally exact representation for the total energy of the three-body system can be interpreted physically in an appealing way. Let

$$P(i) = \&(i) - V(i) \quad (7a) \qquad \text{and} \qquad P(ij) = \&(ij) - V(2,ij) \quad (7b)$$

define the apparent perturbation to the isolated one-body and two-body energies, respectively, when the one-body and two-body systems are assumed present in a three-body cluster. Solving equation (7) for $\&(i)$ and $\&(ij)$, substituting these into equation (4), and then using the resulting $V(1,2,3)$ in equations (3)-(5), leads directly to the following expression for the three-body stabilization energy,

$$E(s,3) = \sum_i P(i) + \sum_{i<j} P(ij) + \&(ijk) \tag{8}$$

This result, based on BEA, provides an operational scheme for the analysis of the pairwise additivity approximation for a three-body system.

When the fluorine is on the x-axis, then the F^--H_2O two-body cluster has C_{2v} symmetry (the x-axis is the main axis for this point group); therefore, any position chosen for the lithium nucleus with $\pm x$, $\pm y$, and $+z$ coordinates is equivalent to the lithium position with the same value for the $\pm x$ and $\pm y$ coordinates, but with opposite sign for the z coordinate. Equivalently, when the lithium nucleus is on the x-axis with $y=z=0.0$, any position for the fluorine nucleus (with $z=0.0$) has a symmetry-generated equivalent position. Other symmetry rules can be generated in a similar manner; in this way, the 250 points computed sample the space at about 600 different positions.

A significant simplification to the problem of a proper sampling of the many-dimensional surface of the $H_2O-Li-F$ complex is achieved not by symmetry or other topological considerations alone, however, but by common chemical sense. Firstly, we note that the Li-F Hartree-Fock binding energy is about -180 kcal/mole, that the Li^+-H_2O Hartree-Fock binding energy is about -35 kcal/mole, and that the F^--H_2O Hartree-Fock binding energy is about -18 kcal/mole, on the average. Hence, the Li-F system has a binding energy about 5 to 6 times greater than the other systems, and hence the energetics of the Li-F system will dominate the complex. If we wish to use a very simple viewpoint, we could say that the $H_2O-Li-F$ system is really no more than a perturbed Li-F system, the perturbation being written as $Li^+-(F-H_2O)^-$ or $(Li-H_2O)^+-F^-$, rather than Li^+-F^-. The water acts as a perturbation, though, of course, a non negligible one.

In Table 2 we report a few of the Hartree-Fock total energy for the 250 configurations selected. In this table, the first column is an identification number for the geometry; this identification number is also used in Table 3. Columns 2-4 report the Cartesian coordinates (in a.u.) for the Li^+ ion; the following three columns 5-7 report the Cartesian coordinates for the F^- ion. The coordinates of the water molecule are held constant; the oxygen nucleus is at the origin of the Cartesian system, the first hydrogen is situated at x= -1.1025738 a.u., $y = 1.4335318$ a.u., and $z=0.0$ a.u., and the second hydrogen is situated at x=1.1025738 a.u., y=-1.4335318 a.u., and $z=0.0$ a.u.

The eighth column of Table 3 gives the total Hartree-Fock energy in a.u. The ninth column gives the stabilization energy (in kcal/mole) of the $H_2O-Li-F$ complex relative to the water molecule [$V(H_2O)$= -76.0552619 a.u.], the fluorine ion [$V(F^-)$= -99.4582356 a.u.], and the lithium ion [$V(Li^+)$= -7.236346 a.u.]. The last five columns report in kcal/mole the stabilization energies for the Li-F molecule, the H_2O-Li^+ complex, the H_2O-F^- complex, the sum of these quantities [i.e., $E(s,2)$], and the difference between the total Hartree-Fock stabilization energy $E(s,tot)$ and $E(s,2)$, namely, $E(s,3)$.

Although, for the most part, the magnitude of the energy $E(s,3)$ is small relative to the total stabilization energy, there are cases where it is comparable to the water-water stabilization energy (5 ± 1 kcal/mole).

TABLE 2 – Cartesian coordinates and Hartree-Fock energy quantities for the Li-F-H₂O complex

Point	Cartesian coordinates (a.u.)						Total Hartree-Fock energy(a.u.)	LiFH₂O	Hartree-Fock binding energy (kcal/mole)				
	X(Li)	Y(Li)	Z(Li)	X(F)	Y(F)	Z(F)			LiF	LiH₂O	FH₂O	Sum of two-body	Difference
5	3.580	0.0	0.0	-5.150	0.0	0.0	-182.9632595	-133.917	-72.431	-36.302	-16.554	-125.288	-8.629
6	2.531	2.531	0.0	-5.150	0.0	0.0	-182.9505370	-125.933	-78.403	-24.402	-16.554	-119.359	-8.574
12	-5.303	5.303	0.0	-5.150	0.0	0.0	-182.9528046	-127.356	-121.174	7.844	-16.554	-129.883	2.527
13	-7.500	0.0	0.0	-5.150	0.0	0.0	-182.9475583	-149.164	-144.984	8.354	-16.554	-153.184	4.020
18	-15.000	0.0	0.0	-5.150	0.0	0.0	-182.8732456	-77.433	-63.957	2.305	-16.554	-78.146	0.713
20	3.580	0.0	0.0	-7.500	0.0	0.0	-182.9203756	-107.007	-56.708	-36.302	-9.860	-102.871	-4.136
21	2.531	2.531	0.0	-7.500	0.0	0.0	-182.9065085	-98.343	-60.817	-24.402	-9.860	-95.079	-3.264
34	3.580	0.0	0.0	-15.000	0.0	0.0	-182.8660829	-73.503	-33.802	-36.302	-2.433	-72.537	-0.965
54	0.0	5.303	5.303	-5.150	0.0	0.0	-182.8887219	-87.144	-69.406	-0.959	-16.554	-86.919	-0.225
55	-3.750	3.750	5.303	-5.150	0.0	0.0	-182.9218686	-107.944	-96.204	3.173	-16.554	-109.585	1.642
56	-5.303	0.0	5.303	-5.150	0.0	0.0	-182.9584424	-131.145	-121.174	4.176	-16.554	-133.552	2.407
71	-5.303	0.0	3.580	-7.500	0.0	0.0	-182.9357134	-116.631	-111.888	4.176	-9.860	-117.572	0.941
96	0.0	0.0	3.580	-7.500	0.0	0.0	-182.9173311	-105.097	-76.222	-18.393	-9.860	-104.476	-0.621
119	5.303	0.0	5.303	5.150	0.0	0.0	-182.9255333	-110.243	-121.174	-8.468	18.095	-111.548	1.304
120	3.750	3.750	5.303	5.150	0.0	0.0	-182.8822694	-83.095	-96.204	-6.200	18.095	-84.310	1.215
138	5.303	0.0	5.303	7.500	0.0	0.0	-182.9260137	-110.545	-111.888	-8.408	8.751	-111.605	1.060
139	3.750	3.750	5.303	7.500	0.0	0.0	-182.8796575	-81.456	-84.805	-6.200	8.751	-82.254	0.708
153	5.303	5.303	0.0	5.150	0.0	0.0	-182.9227519	-108.498	-121.174	-6.541	18.095	-109.620	1.122
169	5.303	5.303	0.0	7.500	0.0	0.0	-182.9230649	-104.694	-111.888	-6.541	8.751	-104.678	0.084
172	-7.500	0.0	0.0	7.500	0.0	0.0	-182.7912930	-26.007	-41.797	8.354	8.751	-24.692	-1.315
191	7.500	0.0	0.0	0.0	7.500	0.0	-182.8657873	-72.753	-59.290	-9.971	-3.386	-72.647	-0.106
195	0.0	15.000	0.0	0.0	7.500	0.0	-182.8889914	-87.313	-84.805	0.291	-3.386	-87.401	0.588
200	7.500	0.0	0.0	-3.140	4.082	0.0	-182.8919270	-89.155	-55.110	-9.971	-21.554	-86.634	-2.521
201	5.303	5.303	0.0	-3.140	4.082	0.0	-182.9120130	-101.759	-74.182	-6.541	-21.554	-102.277	0.518
202	0.0	7.500	0.0	-3.140	4.082	0.0	-182.9935466	-152.922	-138.014	1.736	-21.554	-157.832	4.910
205	0.0	15.000	0.0	-3.140	4.082	0.0	-182.8701770	-75.507	-55.286	0.291	-21.554	-76.549	1.042
207	3.580	0.0	0.0	-4.572	5.945	0.0	-182.9269126	-111.109	-62.398	-36.302	-9.506	-110.206	-2.903
212	0.0	7.500	0.0	-4.572	5.945	0.0	-182.9711405	-138.862	-132.894	1.736	-9.506	-140.663	1.801
217	3.580	0.0	0.0	-4.020	2.275	0.0	-182.9715076	-139.092	-74.372	-36.302	-19.551	-130.225	-8.868
219	0.0	3.580	0.0	-4.020	2.275	0.0	-182.9513267	-126.429	-133.657	20.837	-19.551	-132.370	5.942
220	7.500	0.0	0.0	-4.020	2.275	0.0	-182.8827833	-83.418	-50.876	-0.971	-19.551	-80.397	-3.021
221	5.303	5.303	0.0	-4.020	2.275	0.0	-182.8899024	-87.885	-60.641	-6.541	-19.551	-86.733	-1.152
222	0.0	7.500	0.0	-4.020	2.275	0.0	-182.9206843	-107.201	-91.469	1.736	-19.551	-109.283	2.083

TABLE 3 – Selected geometries for the bond energy analyses (all data in kcal/mole)

Point	E(Li)	E(F)	E(H_2O)	E(LiF)	E(LiH_2O)	E(FH_2O)	E(LiFH_2O)	Total Hartree-Fock energy (a.u.)
5	-17.863	0.802	15.824	-70.180	-60.842	-10.758	9.102	-182.9632595
6	-15.339	0.077	5.528	-76.410	-43.447	-4.646	8.307	-182.9505370
12	-2.227	4.023	-18.734	-130.873	5.171	10.329	4.957	-182.9528046
13	-171.416	60.872	-8.103	-43.224	3.636	6.152	2.920	-182.9875583
18	-0.098	-4.814	-21.081	-66.271	0.771	10.176	3.680	-182.8732456
20	-17.373	0.948	31.375	-56.000	-51.170	-18.673	3.888	-182.9203756
21	-14.997	1.146	19.372	-60.337	-29.601	-15.731	1.907	-182.9065685
34	-18.144	0.016	22.173	-33.191	-40.289	-5.534	1.467	-182.8669829
54	-0.364	-2.848	-16.253	-70.976	-8.729	0.558	5.470	-182.8887219
55	-0.857	-1.015	-17.269	-100.797	-1.932	7.912	6.017	-182.9218686
56	-2.251	3.093	-16.912	-130.459	0.984	8.750	5.651	-182.9588424
71	-1.849	8.458	0.261	-118.371	3.906	-8.605	-0.430	-182.9357134
96	-20.098	1.821	41.655	-75.266	-40.634	-18.999	0.427	-182.9173311
119	-2.970	5.480	-12.337	-133.458	-15.681	38.211	10.511	-182.9255333
120	-1.481	-0.073	-12.540	-102.944	-12.809	37.767	8.987	-182.8822694
138	-1.851	8.339	2.421	-118.508	-10.961	8.015	2.002	-182.9260137
139	-0.820	2.746	2.149	-86.832	-8.377	8.299	1.390	-182.8796575
153	-2.083	5.968	-13.341	-133.843	-12.099	38.143	0.659	-182.9227519
160	-1.851	8.529	0.966	-118.735	-7.570	8.356	1.611	-182.9230649
172	-0.172	0.057	4.447	-41.776	4.121	6.656	0.661	-182.7912930
191	-0.155	0.700	3.756	-59.686	-14.085	-4.153	0.873	-182.8657873
195	-0.054	2.593	-0.681	-86.852	0.506	-2.367	0.143	-182.8889014
200	-0.260	6.849	-8.465	-52.241	-22.059	-14.914	1.936	-182.8919270
201	-0.234	5.197	-13.531	-71.262	-15.217	-9.142	2.431	-182.9120130
202	-3.961	10.699	-19.467	-144.039	-1.558	1.300	4.106	-182.9935466
205	-0.024	2.651	-19.704	-54.539	-1.467	-4.069	1.646	-182.8701770
207	-18.099	1.012	28.348	-61.395	-47.456	-17.519	3.402	-182.9269126
212	-3.681	13.219	-1.102	-142.186	2.125	-0.574	0.660	-182.9711405
217	-17.940	5.852	21.752	-68.827	-64.437	-22.828	7.337	-182.9715076
219	-26.217	14.033	6.276	-132.603	11.118	-1.807	2.773	-182.9513267
220	-0.333	1.471	-10.920	-50.189	-22.401	-4.063	3.039	-182.8827833
221	-0.218	1.001	-15.145	-60.003	-15.730	-1.321	3.532	-182.8999024
222	-0.556	1.792	-19.504	-92.802	-2.779	3.002	3.048	-182.9200543

Thus, in the study of the Li$^+$ and F$^-$ pair in water, the three-body correction E(s,3) might be of importance, and cannot be neglected, without a critical study.

In Table, 3 we report the results of the BEA study. The number appearing in the first column of this table is an identification number for the geometry of the complex (see Table 2). All energies in the table are reported in kcal/mole, except the total Hartree-Fock energy, which is given in a.u. The quantities E(Li$^+$), E(F$^-$), and E(H$_2$O) are just the P(Li$^+$), P(F$^-$), and P(H$_2$O) defined in equation (7a); these energies represent the full perturbation of the one-body energies in the complex. The quantities E(Li$-$F), E(Li$-$H$_2$O), and E(F$-$H$_2$O) are the &(ij) terms from BEA; that is, they represent the two-body interactions in the complex. The term E(Li$-$F$-$H$_2$O) is the "nonclassical" BEA &(ijk) three-body term; hence, the total stabilization energy is easily calculated from Table 3 as E(s, tot) =E(Li$^+$) + E(F$^-$) + E(H$_2$O) + E(Li$-$F) + E(Li$-$H$_2$O) + E(F$-$H$_2$O) + E(Li$-$F$-$H$_2$O)

The analysis of Table 2 (the classical decomposition of the total energy) is very simple. In cases 5, 13, and 18, the fluorine ion is at x= -5.15 a.u. and the lithium ion is at x=3.58 a.u. from the oxygen (i.e., the configuration F$-$H$_2$O$-$Li), or at x= -7.5 a.u., or at x= -15.0 a.u. The two-body interaction for Li$-$F namely V(Li$-$F), and the remaining two-body interactions, V(Li$-$H$_2$O) and V(F$-$H$_2$O), are exactly as described previously. The contribution E(s,3) is simply the difference between the total Hartree-Fock stabilization energy and the sum of the two-body terms. In case 5, the complex F$-$H$_2$O$-$Li is compact and the energy E(s,3) is found to be attractive and large (about -8.6 kcal/mole). In case 18, the Li$-$F$-$H$_2$O complex is very diffuse [R(Li$-$R)= 9.85 a.u. and R(F$-$O)= 5.15 a.u.] and E(s,3) is repulsive and small. We might expect these cases to be typical of Li$^+$-(F$-$H$_2$O)$^-$, but from the data available we cannot confirm this simple interpretation.

Indeed, it is difficult to obtain much physical insight from this part of the table, and perhaps one can only conclude that E(s,3) is about 10% of E(s,2) for a complex which is compact, but tends to decrease the more the complex is separated either into three distant single bodies, or into a two-body pair with a distant third body.

Let us now comment briefly on the BEA data presented in Table 3. Firstly, we note that essentially all the three-body terms are positive (repulsive). Secondly, all the E(Li) terms are negative (attractive) and at times rather large. The largest value of E(Li) found corresponds to case 13 for the Li$-$F$-$H$_2$O complex. The main result that we can obtain from the data in Table 3 is an alternative viewpoint in the binding mechanism of the complex. Having noted that E(s,2) is not too different from E(s,2)+E(s,3), one could deduce that the binding mechanism describing the 3 two-bodies is essentially an adequate description for the binding mechanism of the three-body system. On the other hand, one might suspect that near an equilibrium geometry the total binding energy is the result of a complex mechanism, such as a cancellation of terms. This cancellation of terms seems to be suggested by the BEA data.

From a thorough analysis of the data presented in Table 3, we cannot only conclude qualitatively that the assumption of pairwise additivity is only an approximation, but more importantly, we have provided numerical data on its limitation and/or validity.

3.4 A more Accurate Determination of the Coordination Number of Ions

In this section we report some preliminary data on a study in progress[16] for an accurate determination of the coordination numbers for the ions Li$^+$, Na$^+$, K$^+$, F$^-$ and Cl$^-$. In order to avoid the problem of three-body effects (see previous section), we consider only a single ion rather than an ion-pair surrounded by two hundreds molecules of water; secondly, we are using the latest and more accurate water-water potential obtained from configuration interaction studies[4].

From preliminary data on all the above ions, it is clear that we can now obtain not only the first and second shells radii and coordination numbers, but also those of the third shell. For example, the coordination number for Li⁺ are 4±0.1, 10.3±4 and 18±6; the corresponding radii for the first three shells are at 2.65 A, 4.95±0.2 A and 7.2±0.3 A considering the hydrogen distribution, 1.99±0.05 A 4.0±0.2 A and 6.9±0.3 A considering the oxygen distribution. These data are not final, since obtained only after 5×10^5 random motions of the water. However, from the analyses of the intermediate data, it seems that the above values for the first two shells have converged to the final value (that will require nearly as many additional random motions are those presently computed).

We conclude this section on a rather optimistic note, originated from the consideration that in the last few years not only the techniques both from quantum mechanics and statistical mechanics have reached the point to study the above problems, but we are now in the application stage, as above indicated, whereby one can consider many solvation shells, rather than only one as traditionally done.

3.5 Solvation Potentials for Amino-acids

It is well known that the amino-acids constitute essential building blocks of many macromolecules of biological interest. It is equally well known that most biological molecules interact with each other in aqueous solution at temperatures near room temperature. A problem of interest in this context is the determination of water's structural organization around macromolecules.

Our technique can be summarized into three consecutive steps. First, we compute the interactions of a selected molecule, M, with a molecule of water placed in a sufficiently large number of positions and orientations relative to M, so as to give a reasonable sampling of the potential surface representing the interactions of the "M-water" complex.

In the second step we fit some relatively simple analytical expression to the interaction energy of M with water, computed in the first step.

The third step makes use of the analytical potentials, previously obtained, and determines the structural organization of many molecules of water (representing the solvent) around M for some selected temperature, using Monte Carlo techniques.

Clearly, there are many alternative ways to reach the same goal, namely the determination of the structural organization of water as solvent, around M, for example, one could obtain the structural organization of water molecules around M by direct experimentation. .To our knowledge, however, such approach can be brought to its conclusion in a shorter time and in a more definite manner, if theoretical data, like those here reported, would be available. In addition, the experimental difficulties in obtaining the structural organization of water around M either by X-ray or neutron diffraction (at room temperature, or nearby) are nearly as remarkable as those one encounters in solving the previously explained three steps.

In this section we discuss our results[17] for a selected group of amino-acids, namely cistine, cysteina, methionine, glycine, alanine, valine, leucine and isoleucine, arginine, asparagine, lysine, glutamine, glutamic acid, aspartic acid, proline, hydroxyproline, tryptophan, tyrosine, threonine and serine.

The computations we shall discuss below are not as accurate as those performed for the water-water interaction, the reason being that for the case of water our work aimed at a duplication and verification, via simulation, of accurate X-ray and neutron diffraction experiments. For the case of macromolecules in water, the situation is qualitatively very different; indeed, only macromolecules in crystals and not in solution can now be tackled experimentally, by diffraction techniques. Our computations are likely the needed ground work to obtain structural determinations with about 0.3 A resolution. At the first step (previously described) the computation is definitely superior and could yield structural data with resolution in the range 0.2-0.1 A; however, the errors introduced in step two and later in step three will decrease our reliability to the above given limit. The interactions for the amino-acids and one molecule of water are computed in the SCF—LCAO—MO approximation. The molecular orbitals are expanded in a linear combination of contracted gaussian functions, each one centered at a nucleus of the system[18]; in turn, the contracted gaussian functions are a prefabricated (on the base of atomic computation) linear combination of primitive gaussian functions[18], We have used a 7/3 basis set. The molecule M is considered fixed in space and the water molecule is placed at different distances around M. For a given position of the oxygen atom of H_2O, one or more orientations for the water molecule are considered ("orientations" in this context is used to indicate that the hydrogen atoms of H_2O might point towards, away or otherwise relative to M). For each position and orientation of the water molecule we have computed the total energy of the system $M+H_2O$, designated as E(M,W); by subtracting from E(M,W) the energy of the water molecule, E(W), at infinite distance from M and the energy of the molecule M, E(M), at infinite distance from the water molecule, we obtain the interaction energy, I(M, W) between M and water, i.e. I(M, W)=(E(M, W)−E(M)−E(W).

For each case we have computed the net charges, (NCH) and the Molecular Orbital Valency State (MOVS) previously discussed[19,20]. The second quantity, MOVS, represents the energy of an atom, when in the molecule M, relative to its energy when in the atomic ground state; therefore the MOVS constitutes an index of the energetic variations that follow the formation of a molecule from separated atoms[21].

The NCH and MOVS are clearly related to the hybridization of the atoms in $M^{[21]}$; both quantities can be computed in a fast, simple way, once an SCF−LCAO−MO solution has been obtained. For the water molecule we have taken a standard geometry and the basis set chosen yields a value of E(W)= -75.733058 a.u. For the amino-acids analyzed, we have computed a total of 2212 different positions or orientations of the water molecule.

As stated in the first part of this section, the fitting of the discrete sampling of points is intended as input to the Monte Carlo simulation, that in turn will yield the probability distribution of many molecules of water surrounding M at a pre-determined temperature. In the Monte Carlo procedure many configurations of the water molecules (corresponding to different positions and/or orientations) are randomly chosen and statistically weighted; therefore, it is essential to use a simple and computationally fast analytical form to fit the interaction energy, since for each configuration it is required to recompute the quantity E(M, W), in addition to the water-water interaction.

We have decided to use a Lennard-Jones type potential. We note that this potential has been used by several authors in describing intermolecular interactions for many biological systems including amino-acids[22, 23]. The interaction energy is fitted by the following expression

$$E(M, W) = E(M) + E(W) + I(M, W) \tag{9}$$

where

$$I(M,W) = \sum_i \sum_j I_{ij}^{ab}(M,W) = \sum_i \sum_{i \neq j} (-A_{ij}^{ab}/r_{ij}^6 + B_{ij}^{ab}/r_{ij}^{12} + C_{ij}^{ab} q_i q_j /r_{ij})$$

and where A_{ij}, B_{ij}, C_{ij} are fitting constants, r_{ij} is the distance between an atom, i, in M and an atom, j, in the water molecule, q_i and q_j are the net charges (NCH) for the atoms i and j, a and b are indices that not only differentiate between atoms (e.g., differentiate a hydrogen atom from a carbon atom), but also, within a group of atoms of equal atomic number z, differentiate its environmental conditions in the molecule, according to some quantitative criteria, obtained directly from the SCF−MO study of M and water. We note that physically the r^{-12} term represents the repulsion for the atoms i and j when they are close to each other, the r^{-6} term, the attractive interaction at intermediate regions, the r^{-1} term, the point-charge, point-charge interaction, attractive or repulsive depending on the sign of q_i and q_j (C_{ij} is always positive). We have made use of the values of MOVS as the criterion, to select the classes identified by the index a.

On this basis we have differentiated the hydrogen atoms in 4 classes: 1) an hydrogen in the $-NH_2$ fragment; 2) a hydrogen in the $-(CH_2)-$ or in the $-(CH)$ fragments; 3) a hydrogen in the $-CH_3$ fragment and 4) a hydrogen in the $-OH$ fragment. For the carbon atoms we differentiate into the following classes: 1) a carbon in the -COOH fragment; 2) a carbon in the -CH$_3$ fragment; 3) a carbon in the -(CH$_2$)- or in the aliphatic -(CH)= fragments; 4) a carbon in the -(CHR)- fragment of the glycine backbone. For the oxygen atom there are two classes: one for -OH and the second for C=O.

This classification was sufficient to describe glycine, alanine, valine, leucine and isoleucine, however, the MOVS's values for the nitrogen atoms in arginine, asparagine, lysine, glutamine, aspartic acid and glutamic acid pointed out the need to extend the classification for the nitrogen in -NH$_2$ for H$_2$N—CHR—COOH, a class for nitrogens in =NH for HN=C (R(1) R(2)), and a class for nitrogen in -NH$_2$ for H$_2$N—CR=NH. For proline, hydroxyproline and hystidine the above classification was not sufficient and we introduced two new classes for carbon, one in the hystidine ring and the other in the C—OH group of hydroxyproline; one new class for nitrogen in N—H group of hystidine ring and another one for hydrogen in the ring.

Equivalently for tryptophan, tyrosine, threonine and serine we introduced three additional classes for the carbon atoms of phenyl radical, one of them describing the general atom and the other two the carbons connecting the two rings in triptophan.

For the sulphur containing amino-acids, we need only three additional classes, one for the sulphur atoms and the second for hydrogen atoms in -SH, and one for carbon atoms bonded to sulphur.

In total we have therefore considered 6 classes for the hydrogen atoms, 10 for the carbon atoms, 2 for the oxygen atoms, 4 for the nitrogen atoms, and 1 for the sulphur atoms. Correspondingly we have computed twenty-three sets of interaction constants with the oxygen of H$_2$O and twenty-three sets of interactions with the hydrogen of H$_2$O.

The standard deviation (between the 2212 total energies computed in the SCF—MO approximation and those fitted) is 0.605 kcal/mole. This done, we have considered the phenilalanina as a test case[24], to see if the obtained fitting constants are sufficiently transferable to describe a molecule interacting with water if composed of atoms that experience the same field (due to the nearest neighbours) as previously analyzed in different amino-acids. This test indicated that the derived fitting constants yield a potential that is in remarkable agreement with quantum mechanical computation (standard deviation of about one half kcal/mole).

We have then used the fitting constant on a number of threeglycines (each one at a different conformation) interacting with water[25]. We should remind that the -NH$_2$ group of the amino-acids is replaced by the -NH- group in the threeglycines and that the -COOH group is replaced by a -CO- group. The standard deviation between the computed interaction energies of threeglycines and water using SCF—LCAO—MO techniques and those obtained using the fitting constants is of about 2 kcal/mole. From the work on the threeglycines with water we have obtained new fitting constants to describe the -NH- and the -CO- groups, namely the backbone of proteins[25].

Using the same basis set previously described (7/3), computations have been carried out for the four bases of DNA[26] and other computations are now being completed for the sugar-phosphate-sugar complex[27]. To conclude, by now we have a rather large set of constants that allows to describe the interaction of water with any protein and with DNA. These constants allow a) to obtain the interaction energy with water in a matter of milliseconds of computer time; b) are the needed input for Monte Carlo computations, to obtain statistical data on distribution of many molecules of water around biological molecules (for the amino-acids we restrict ourselves to the use of two hundreds molecules of water).

In Fig. 6 we report the energy contours (spaced by 1 kcal/mole) of the interaction energy of water with the amino-acid phenilalanine and with the guanine base. The selected plane is the one containing the aromatic group. The area examined is of about 20 x 30 A. The contours join iso-energetic points, namely points with the same interaction energy between the molecule and water; each point is obtained by fixing the oxygen atom of H$_2$O in the plane at a grid point of 1 A interval and by rotating the hydrogens in any direction such as to find the energy minimum for the interaction with the molecule in study.

Such diagrams can be obtained in a very short computer time and the two reported in Fig. 6 are two out of many we have analyzed.

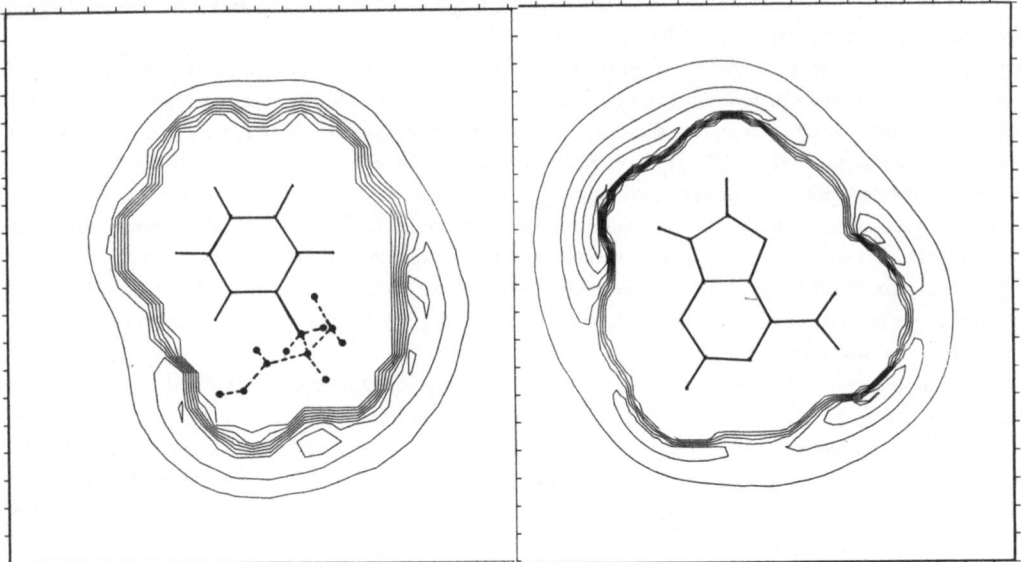

Fig. 6   Interaction of water with phenilalanine and with adenine. The interval between the energy contours is of 1 kcal/mole. The marks on the borders of the figure are given at 1.5 a.u. intervals of length. The selected planes are those of the phenil group and of the skeleton of adenine, respectively. Repulsive energies higher than 3 kcal/mole are not reported. In phenilalanine (on the left) the dashed bonds refer to those atoms that are not in the plane of the phenil group. (See Fig. 5 Part 2 for the equivalent study in the case of water-water interaction)

3.6 Conclusions

The study of the amino-acids here described and the parallel study of the interaction of water with the 4 DNA's bases, with the sugar-phosphate-sugar, and with the polyglycines[26] constitute the ground work to initiate an analysis on the solvation of proteins. A computer program applied to the lisozime interacting with water has been prepared[28] and now is in use (it consists essentially of the same program that has been used to obtain Fig. 6.)At the same time, we are constructing the interaction potentials for a number of amino-acids in the ionic form[29].

We would therefore like to conclude as follows: there is a large amount of very recent work that, with different degrees of accuracy, is directed toward the solvation study of biological systems. Sometimes one or more of the following approximations are introduced: 1) often one uses a 4/3 gaussian basis set that is known to give incorrect position for the energy minima; 2) the number of points and orientations (for the water interacting with a given molecule) analyzed with quantum mechanical methods are often too few to provide correct conclusions; 3) temperature effects are often ignored and therefore no statistical distribution can be considered; 4) water-water effects are often ignored; 5) the pairwise approximation is often used without careful consideration of three-body effects.

Likely, with time, a more critical attitude will certainly become more general. However, what seems very clear is that the problem of solvation for even complex molecules is being pursued to a detail that a few years ago was generally considered totally   unfeasible .

3.7 References

1 J.O. Bockris and P.P.S. Saluja, j' Chem. Phys. 76, 2140 (1972).
2 A. Pullman and B. Pullman, Quarterly Review of Biophysics 7, 4 (1975).
3 E. Clementi, R. Borsotti, J. Fromm, R.O. Watts, Theor. Chim. Acta (in press).
4 J. Fromm, E. Clementi and R.O. Watts, J. Chem. Phys. 62, 1388 (1975); R.O. Watts, E. Clementi and J. Fromm, J. Chem. Phys. 61, 2250 (1974).
5 H. Popkie, H. Kistenmacher and E. Clementi, J. Chem. Phys. 59, 1325 (1973); H. Kistenmacher, G.C. Lie, H. Popkie and E. Clementi, J. Chem. Phys. 61, 549 (1974).
6 Private communications, W. Kolos, Department of Theoretical Chemistry, University of Warsaw, W. Pasteura 1, Warsaw, Poland.
7 J.W. Kress, E. Clementi, J.J. Kozak and M.E. Schwartz, J. Chem. Phys. 63, 3907 (1975).
8 J.A. Barker and R.O. Watts, Chem. Phys. Letters 3, 144 (1971); R. Barsotti "Documentation of a Monte Carlo program for use in statistical mechanics problems" Technical Report n. 13/75, Istituto G. Donegani, Soc. Montedison, Novara-Italy.
9 H. Ulich, Z. Elektrochem. 36, 497 (1930).
10 E. Darmois, J. Phys. Radium, 8, 177 (1942).
11 J. Stuehr and E. Yeager, "Physical Acoustics", Vol. II part A, W. P. Mason, Ed. Academic Press, New York, chapter 6 (1965).
12 S. Broersma, J. Chem. Phys. 27, 481 (1957).
13 R.W. Creekmore and C.N. Reilly, J. Chem. Phys. 73, 1563 (1969).
14 B.P. Fabrigand, S.S. Goldberg, R. Leifer and S.G. Ungar, Mol. Phys. 7, 425 (1964).
15 See part 1 and reference thereby given.
16 E. Clementi and R. Barsotti (unpublished data).
17 O. Matsuoka, M. Yoshimine and E. Clementi, J. Chem. Phys. 64, 1351 (1976).
18 This work has been presented at the International Symposium of Theoretical Chemistry, Boulder, Colorado (June 1975); VI Simposio Chimici Teorici di Lingua Latina, Arles (France) Sept. 1975. The entire set of papers is now in press (see Part 1, reference 32).
19 E. Clementi and D.R. Davies, J. of Comp. Phys. 1, 223 (1966).
20 R.S. Mulliken, J. Chem. Phys. 23, 1883-1841-2338-2343 (1955).
21 E. Clementi, J. Chem. Phys. 46, 3842 (1967); E. Clementi and A. Routh, Int. Journ. Quantum Chem. Vol. VI, 525 (1972); H. Popkie and E. Clementi, J. Chem. Phys. 57, 4870 (1972); see, in addition the detailed discussion of Part 1 of this work.
22 See, for example, A. Momany, R. F. McQuire, A.W. Burgess and H.A. Scheraga, J. Chem. Phys. 79, 2361 (1975); ibid 78, 1595 (1974); L.L. Shipman, A.W. Burgess and H.A. Scheraga, Proc. Nat. Acad. Sci. USA 72, 543 (1975); ibid 72, 854 (1975).
23 W.K. Olson and P.J. Flory, Biopolymers 11, 1 (1972); ibid 11, 25 (1972); ibid 11, 57 (1972).
24 Unpublished data from E. Bolis and E. Clementi.
25 Unpublished data from D. Ferro, R. Scardamaglia and E. Clementi.
26 Unpublished data from R. Scardamaglia and E. Clementi.
27 Work in progress by J. Fromm. C. Tosi and E. Clementi. This work is a continuation of a study by O. Matsuoka, C. Tosi and E. Clementi on the conformational analyses of the sugar-phosphate-sugar complex (to be submitted).
28 Work in progress by A. Jonath, G. Ranghino, D. Ferro and E. Clementi.
29 Unpublished data from A. Petrangelo, R. Scerdamaglia and E. Clementi. Such data refer to few amino-acids in the zwitter-ion form. Additional studies are now in progress (by L. Barino, R. Scardamaglia and E. Clementi) for the aspartic acid and the glutammic acid in the ionic form when part of a protein.